We Culture

12 Skills for Growing Teams in the Future of Work

Luciana Paulise

QUALITY PRESS

Milwaukee, Wisconsin

We Culture: 12 Skills for Growing Teams in the Future of Work
Luciana Paulise

American Society for Quality, Quality Press, Milwaukee 53203
All rights reserved. Published 2022
© 2022 by Luciana Paulise

Publisher's Cataloging-in-Publication data

Names: Paulis, Luciana, author.
Title: We culture: 12 skills for growing teams in the future of work / by Luciana Paulise.
Description: Includes bibliographical references. | Milwaukee, WI: Quality Press, 2022.
Identifiers: LCCN: 2022935206 | ISBN: 978-1-63694-017-5 (paperback) | 978-1-63694-018-2 (epub)
Subjects: LCSH Leadership. | Corporate culture. | Organizational change. | Business communication. | BISAC BUSINESS & ECONOMICS / Business Communication / General | BUSINESS & ECONOMICS / Leadership | BUSINESS & ECONOMICS / Mentoring & Coaching
Classification: LCC HD58.8 .P38 2022 | DDC 658.4/06—dc23

ASQ advances individual, organizational, and community excellence worldwide through learning, quality improvement, and knowledge exchange.

Bookstores, wholesalers, schools, libraries, businesses, and organizations: Quality Press books are available at quantity discounts for bulk purchases for business, trade, or educational uses. For more information, please contact Quality Press at 800-248-1946 or ask@asq.org.

To place orders or browse the selection of all Quality Press titles, visit our website at: http://www.asq.org/quality-press

Printed in the United States of America.

27 26 25 24 23 LS 7 6 5 4 3

QUALITY PRESS
Quality Press
600 N. Plankinton Ave.
Milwaukee, WI 53203-2914
Email: books@asq.org
Excellence Through Quality™

Contents

"A *We Culture* is a culture in which anyone can 'stop the line' if they sense something is not right."

— Luciana Paulise

Foreword

· ·

B eginning in the late 1980s, Dr. W. Edwards Deming used the term "the prevailing style of management" to describe the administration style of organizations that are characterized by directives that foster, if not promote, suboptimization, wherein what's good for the organization is subverted by what's good for the elements of the system. In his last book, *The New Economics,* published in 1993, he proposed "The prevailing style of management must undergo transformation. A system cannot understand itself. The transformation requires a view from outside." In this book, Luciana Paulise, a devoted student of the Deming philosophy, provides an illuminating outside view for newcomers to organizational dynamics, those who are open to thinking beyond constraints, which include the mythical pursuit of perfection. She does so through the simplicity of a binary model of organizations, one that contrasts the thriving interdependent thinking of a *We Culture* with the intense independent thinking of a *Me Culture.* Page by page, she couples what she's learned from Deming's theory of management with her diligent personal research into organizations many have heard of and others she's uncovered in her quest to reveal evidence of the dominance of the prevailing style of management and the escalating emergence of We Cultures.

Using another simple model, organizations dominated by a Me Culture represent "organizations as usual," characterized by symptoms that include the spoken, as well as the unspoken, existence of a "most important part" and a prevalence of blame placed on individuals rather than the greater system in which they operate. A favorite mental picture of the prevailing style of management is a view of the completed individual elements, seemingly independent and floating in close formation, never connected to each other. The management actions (and thinking) that unknowingly sustain such nonsystemic behaviors are driven by an unrecognized and, therefore, unstated, set of beliefs and assumptions that are highlighted in this book. Another tell-tale sign of these beliefs are management practices that focus on parts and ignore, if not underestimate, relationships and interdependencies.

Away from work, while we live in a society that praises teamwork, we also live in a society that falls victim to either rewarding or blaming the fictional last straw, very consistent with a Me Culture. Look no further than news reports to witness blame of the pilot of an airplane crash or the celebration of an athlete in professional sports for a play that appears to win a game. While Deming was widely praised for his transformational efforts in Japan, he also noted that he was lucky to have been invited to contribute.

What Luciana unmasks in this book represents the flexibility, vitality, and sustainability of "organizations as unusual," those in which teamwork is far more than a simple expression, those with an intense and expanding sense of the purpose and relatedness of all parts of a system, from the system representing the products and services they design and deliver to the system in which they operate. In my experience in studying organizations for the past 30 years, "organizations as usual," whether operating in a time of crisis or in a period of respite and reflection, snub documented processes and instead prosper on results. They do so with little regard for unintended consequences for their workforce, suppliers, and customers, or the future of the organization. That is, there is generally little interest in "by what method?" to quote Deming. Yet, without a method, how would "organizations as usual" transform to "organizations as unusual"? Chapter by chapter in this book, Luciana shares her experience in guiding clients to discover the potential energy embedded in the We Culture of "organizations as unusual," thereby enabling leaders at any level to take local actions to counteract the forces of destruction that serve to cripple organizations hindered by Me Cultures.

–Dr. Bill Bellows
President of InThinking Service, Inc.
Los Angeles, California

Preface

· ·

H ave you ever felt like slacking off instead of going to work? Your entire being is unwilling to get out of bed, even if it's just to put on your slippers and sit in front of your laptop in your home office? Or, worse yet, drag yourself out of bed to attend some unproductive meetings?

For a period of time that was me. In today's society, about 80% of workers across the globe[1] are not engaged at work. They daydream about the day when they will be free to do what they love. In many cases, they even love what they do but not how they have to do it.

Why does work have to seem so dreadful? Unfortunately, that's what many workers today are pondering, which is leading to a movement being coined as the "Great Resignation." Fueled by the COVID-19 pandemic, the Great Resignation has workers considering leaving their jobs. The number of workers who are switching their jobs has nearly doubled, and many companies can expect to have 41% of their workforce consider leaving within a year. Disengagement not only costs employees their jobs and stability, but disengaged workers also penalize the global economy. According to a 2021 Gallup survey, lost productivity cost the United States about $8.1 *trillion* (author emphasis added) annually, nearly 10% of the gross domestic product (GDP).[2] But the decline in productivity and increase in disengagement of works can be stopped. It needs to stop, and you can do something about it.

The *We Culture* book is an invitation to co-create a culture in which we all can voice loudly what we like and who we are to fully engage ourselves in the flow of work and ultimately benefit from unlocking each others' strengths. Simultaneously, our companies can produce better products and services, more cost effective and customer oriented, in a work environment that is employee-friendly, productive, and fulfilling.

Whether in-person, remote, or in the metaverse,[3] people can enjoy life and work in a delicate balance where we all benefit and are willing to sustain over time.

In the pre-pandemic era, company culture was a given. Employees were expected to observe and mimic their leaders to follow the unspoken rules. The pandemic, however, revealed that most organizations lacked a corporate culture and operational agility to respond to permanent change.

Today's global, complex, and disruptive business environment demands companies to make their organizations more adaptive and agile. It's imperative that organizations upskill their personnel and set their culture intentionally. The companies that develop this culture right will have an incredible, competitive advantage. Leading change is everybody's job; nobody can do it alone. A systemic and collaborative approach is needed: the *we culture* mentality.

The research in this book is based on what I have read, experienced, and aspired to in nearly a 20-year career. I have been lucky to learn firsthand from the companies where I worked as an employee, a leader, and an external consultant. I also visited and learned from organizations like Google and Zappos, interviewed many business leaders who operate in different countries, and had access to hundreds of W. Edwards Deming's manuscripts from the Library of Congress.

When I visited the Las Vegas, Nevada, headquarters of Zappos (the online shoe and clothing retailer), I confirmed that my ideal of "we" company was no longer an ideal but a goal to achieve. At Zappos, employees realize that teams are the way to accomplish goals, employees' ideas really do matter, and self-organization prevails over the chain of command. Zappos doesn't designate only managers; everyone focuses on their job at hand—not their title or position. From employees to clients to suppliers, everyone is important. Zappos staff are so eager to share what they have learned—another characteristic of we cultures—that their *culture maestros* offer company tours. These culture maestros offer you a guided tour, showing you *everything* from mini-golf stations to their all-hands auditorium and spaces to take a nap. The environment was so inviting that even my seven-year-old daughter still dreams of returning to the Zappos building.

Does that arrangement look impossible or impractical? It may seem so, but it's becoming more of a reality at innovative companies. Is it easy? Probably not, but it is achievable and can generate great bottom-line results at the same time. As Zappos' former chief executive officer (CEO), Tony Hsieh, put it: "Chase the vision, not the money. The money will end up following you."

After working with hundreds of teams and interviewing multiple company leaders, I have found 12 necessary behavioral skills for the future of work to lead, care, and collaborate in a face-to-face, remote, or hybrid workplace. They will start as desired behaviors that all the team members will share and embrace until, with everyday practice, they will become skills.

Some concepts will be new to you; some are becoming more commonly accepted, while other concepts may have been suggested decades ago by great thinkers but they didn't stick.

The 12 behavioral skills are the fundamentals of the CARE model to:

- **Connect** with your employees and teammates to drive innovation.
- **Attend** intentionally to every detail to drive quality.
- **Respect** the different needs and characteristics of the team members to drive engagement.
- **Empower** employees to act in a self-organized way to drive agility.

These skills will help you create a culture of conscious teamwork to increase employee engagement, agility, quality, and innovation.

I've divided the book into four parts with each part focusing on an area of the CARE model and diving deeper into discussing the skills related to that area. At the end of each chapter is a "hands-on" exercise for you to put into action what you've just read. I believe part of applying what you read is exercising your mental muscles in order to develop greater strength in that skill set.

I also invite you to visit my website, www.theweculture.com, and download the We Culture app on your phone, to explore other tools, self-assessments, and online training available. Take this once-in-a-lifetime opportunity to reinvent how you work, and enjoy it.

References

1. Gallup, "State of Global Workplace: 2021 Report," *Gallup*, 2021, https://www.gallup.com/workplace/349484/state-of-the-global-workplace.aspx

2. Gallup, "State of Global Workplace: 2021 Report," *Gallup*, 2021, https://www.gallup.com/workplace/349484/state-of-the-global-workplace.aspx

3. The metaverse is a three-dimensional virtual world where everyone can connect no matter where they are physically. It is facilitated by the use of virtual and augmented reality headsets.

Introduction

· ·

T ony Hsieh, former Zappos CEO, said, "If you get the culture right, most of the other stuff, like great customer service, or passionate employees and customers, will happen on its own naturally." He affirmed that most companies don't focus as much on company culture because the return on investment (ROI) is usually two- to three-years down the line. That may seem like a long time, but changing habits and routines is worth it although not easy to do.

As a quality engineer, my job is to help teams deliver products and services that are more affordable, higher in quality, and more efficient. I have worked with multiple groups across various industries, countries, and company sizes. Yet, I've always noticed a common denominator: the groups' abilities to improve their processes were impacted by certain common behaviors and thoughts that their team members developed over time. Examples of these behaviors were blaming someone when a metric was going south or saying "this is not my job."

Such behaviors and ideas become part of the company's culture. Company culture is the sum of behaviors and ideas, acting like a pair of glasses—through this filter, team members see the company with a certain bias (knowing what is OK). Culture basically guides employees to behave in a similar fashion under similar circumstances to avoid being "left out."

In a small business, owner behavior can impact company culture heavily. Whereas, in a big company, all the leaders' behaviors—from management to the executive team—model the company's culture.

This bias could also be positive, resulting in what I reference throughout this book as the "We Culture"—driving better results, a willingness to innovate, collaboration, and increased engagement. When behaviors are viewed as negative, they can result in what I refer to as a "me culture," preventing the team from achieving results or improvement.

The ROI of a We Culture

Many companies have been successful despite not functioning as a "we company," but that success doesn't mean they have achieved optimal results. In a Me Culture these negative results are normalized and even accepted. The company culture is toxic and doesn't allow its staff or leadership to take results to the next level. In the past (and even still today), toxic company cultures didn't have an impact on the bottom line because goals were still achieved and employees were still getting paid. Current researchers, however, show how toxic cultures can reduce employee engagement and, ultimately, the employee's willingness to stay in the company, which can damage the company's financial and operational results drastically. Productivity is not at the level it could be, and costs related to personnel issues are way too high.

> *We rely on highly skilled personnel and, if we are unable to retain or motivate key personnel, hire qualified personnel, or maintain our corporate culture, we may not be able to grow effectively.*
>
> –Sundar Pichai, Google and Alphabet CEO[4]

First of all, what is engagement? Engagement is willingly committing daily to a set of desired behaviors or company culture. Engaged workers feel like owners of their work and, thus, are able to drive change, innovation, and productivity for the company.

On the other hand, disengaged workers are fundamentally disconnected from their work. They tend to be less productive and report being less loyal to their companies, more stressed, and less secure in their work. They report more missed days of work

(3.5 more days per person annually), more days of work missed for illness (0.55 days per person annually), and are less satisfied with their work-life balance.

According to Gallup research results, lower productivity of actively disengaged workers penalizes U.S. economic performance to the tune of $960 billion and $1.2 trillion per year.[5] Disengaged workers report more missed days of work for illness and increased turnover.[6]

Keeping employees engaged during times of crisis is even more challenging. How can leaders deliver quality goods and services if their workforce is disengaged?

Something the COVID-19 pandemic exposed was that most organizations lacked the culture and operational agility to respond to a permanent change even though they could move quickly to operating remotely. The pandemic accelerated trends that were expected in a few years' time. A recent survey[7] showed that 92% of human resources (HR) leaders set employee experience as a top priority in 2021 because of the remote work challenges and the need to improve retention rates.

A Gallup study[8] shows employee engagement impacts performance outcomes significantly. The median percent differences between top-quartile and bottom-quartile units are:

- 10% in customer loyalty/engagement
- 23% in profitability
- 18% in productivity (sales)
- 14% in productivity (production records and evaluations)
- 18% in turnover for high-turnover organizations (those with more than 40% annualized turnover)
- 43% in turnover for low-turnover organizations (those with 40% or lower annualized turnover)
- 64% in safety incidents (accidents)
- 81% in absenteeism
- 41% in quality (defects)

Table I.1   How employee engagement impacts the bottom line.

The relationship between engagement and organizational results is a high price to organizations. Therefore, it's vital for companies to engage their workforce in order to build a competitive advantage.

In a Boston Consulting Group (BCG) study of 40 digital transformations, companies that focused on culture were five times more likely to achieve breakthrough performance than companies that neglected culture.[9]

You cannot force engagement, but you can make it easier to be engaged than disengaged. Defining an intentional culture makes it easier for team members to become engaged and follow the desired behaviors.

Key Insight

Defining an intentional culture makes it easier for team members to become engaged and follow the desired behaviors.

Why are Employees Disengaged?

So why are employees disengaged? There are many reasons why employees are disengaged, but to understand the "why," you have to get to the root cause. A tool that I love to use often to understand the root cause of a problem is the 5 Whys: define a problem and ask "why" five different ways to find out more about that issue.

For example:

Problem – Employee turnover is higher than ever.

1. Why is employee turnover higher than ever?
   Because when employees are disengaged, they tend to change jobs more often.

2. Why are they disengaged?
   Because they don't feel their leaders care about them.[10]
   They don't feel listened to, respected, or empowered.

3. Why do they feel leaders don't care about them?
   Because there is a disconnect between leaders' and employees' needs. Leaders are under so much pressure that they follow the company culture, but that's not enough.

Their own supervisors demand metrics and results, while team members demand listening, time, and purpose (plus money and results).

4. Why is there a disconnect?
   Because team members and leaders have different objectives to achieve that are not aligned. While employees work for the customer, leaders should work to help the employees, but instead, they work for their supervisors in order to get more benefits in the short term.

5. Why is it that objectives are not aligned?
   Because leaders and employees are working for their own good instead of the greater good, which is what they learned in previous jobs to be more successful. There is no intentional culture or defined set of behaviors to help employees work collaboratively instead of competitively.

Root cause defined: There is a systemic disconnect, because culture has not been set intentionally, and no leader can change independently.

No wonder why, even though companies spend $15 billion[11] annually on leadership development, a study[12] showed that 75% of Americans say their "boss is the most stressful part of their work day."

Employees usually wait for directions; they don't contribute or share their thoughts without asking and usually avoid asking for help. They are afraid of asking. If you think about it, most of the mistakes that are made every day in organizations are due to questions not asked or not answered and issues not raised in a timely fashion. The reason why this happens is that people don't have the right incentives to do so. Even worse, employees receive incentives to do precisely the opposite. In the past, they have been punished for speaking up.

Furthermore, there are incentives for individual sales, employee rankings, school grades, competition, and leading without asking. There are no incentives for contributing, only numerical goals that push the system, are hard to attain, and offer no means to make the system better. Employees then get frustrated.

Employees are now looking for more than just a paycheck. They want their purpose to align with the company's purpose, unleash their strengths, and have more freedom to work independently.

And they want to develop meaningful and psychologically safe relationships with their teammates, especially with their leaders.

Key Insight

Employees are now looking for more than just a paycheck. They want their purpose to align with the company's purpose, unleash their strengths, and have more freedom to work independently.

The Leadership Role

Before the pandemic, leaders had more relevance and were expected to be in charge. Suddenly, the shift to a remote workforce and a more volatile environment poses tremendous challenges for leaders.

The solution is not to fix leaders by pushing them to work harder and hoping they will do better. The root cause of the problem gets fixed when we give leaders and team members a better map to navigate—a more precise understanding of what is expected and the power to set their own journey.

Employees need coaches and mentors who help them build new habits to increase feedback sessions, listen more, and be more like facilitators than directors. To change the culture, all employees—especially leaders—need to learn new skills.

Key Insight

To change the culture, all, and especially leaders, need to learn new skills.

One of the most important skills leaders have to develop is facilitating interactions to succeed in a changing environment. Leaders should no longer be in charge of telling their teams what to do, but they should empower employees to be in charge of themselves. Like in sports, leaders can be outstanding performers, but you need a team to win a championship.

A team is like an orchestra. Everyone is needed to succeed, and everyone is responsible for delivering the best in their role. The director of the orchestra doesn't play for them; he or she makes sure they are aligned and communicating with each other.

Key Insight
One of the most important skills leaders have to develop is facilitating interactions.

In a we culture, less-controlling management styles are required. Team members start trusting more in themselves and expecting less from their leaders. It requires a top-down and bottom-up approach in which everyone in the organization is responsible.

Amy Edmondson, a Harvard researcher, confirms that the team structure and shared beliefs shape team outcomes. It is not just about defining rules but making sure those rules become the expected behaviors and beliefs leaders and team members follow.

Once you have the desired habits ingrained into your employees, you simply need to hire new employees, let them know the behaviors, and make sure they meet other teams and coaches through the support structure. Then, organically, they will look for teammates who want to work together to accomplish a goal aligned to the mission.

The role of the *we culture* is to help companies develop a framework to become more team oriented, guiding everyone to think in terms of how what they do contributes to themselves as well as to the greater good. Instead of putting all the pressure on the leader, a we culture trusts that not only the leader, but also the organizational context and team members, are in charge of the results.

Transforming a culture, whether a team or a whole company, involves identifying the behaviors you would like to see become consistent practices and then instilling the discipline of actually committing to them. Communicating what you want to do is not enough; you need influential people to change their own behaviors first and start an imitation game, because your culture is defined by how people behave.

Key Insight
Your culture is defined by how people behave. Transforming a culture, whether a team or a whole company, is a matter of identifying the behaviors you would like to see become consistent practices and then instilling the discipline of actually committing to them.

For example, when you want your children to do something, they won't always follow what you say; however, they will follow what you do. If you don't want them to watch TV all day, have books around, feed them well, and talk about the importance of studying. You cannot control them all day: you don't know what they do at school or at their friends' houses. You can try to be a great parent, but you can't be there at all times. So, you have to give them the tools so they trust that behaving the same way everywhere every time is the best option for them and the people around them.

The Opportunity to Choose Your Culture

Changing the culture, employees' habits and shared beliefs, is an opportunity for companies to improve employee engagement and profitability. The first step is to define the culture intentionally.

Defining an intentional culture involves positioning people for success by guiding them through the expected behaviors and habits that the company values and respects, which will help them make decisions and achieve results that are aligned with the company's purpose. Habits can make up to 80% of our daily activities, so if companies can predefine the best team habits, communicate them, and design a context that will promote them, it is much easier to model the desired culture.

Key Insight

If companies can predefine the best team habits, communicate them, and design a context that will promote them, it is much easier to model the desired culture.

The purpose is what the company wants to do; branding is what it says it will do, but culture is what it actually does. Engagement is how well you meet employees' needs through your culture to help them perform at their best. When purpose, branding, and culture are aligned, employee engagement increases, performance improves, and the connection between team members and leaders is higher.

Culture is transmitted from employee to employee by example, just by observing each other, as a way to fit in and be accepted.

When hybrid teams are the norm, expected behaviors can be blurred or not as easy to imitate as in a face-to-face team. Employees, especially new hires, can be more confused about what is expected from them. It is hard to grasp what is OK and what is not OK. If rules continue to be unspoken, and now they are also unseen, then culture ends up being different from one employee to another. The way to make decisions, aversion to risk, or disposition to innovate may be totally different within the same team.

Culture can and should be set intentionally so every employee flows in the same direction, following the same rules. The companies that develop these rules right will have an incredible competitive advantage, as when all your employees are enjoying their work and giving their best, they will be able to achieve their self-imposed goals faster. When companies focus on changing the culture, 80% of the behaviors will be aligned with 20% of the effort, following the 80-20 Pareto rule.

Key Insight

Culture can and should be set intentionally so that every employee flows in the same direction, following the same rules.

Pareto Rule: The Pareto rule, also called the 80/20 or the law of the vital few, states that for many outcomes, roughly 80% of consequences come from 20% of causes. This rule will be used on various occasions throughout the book as a means to prioritize problems, tasks, or risks, and focus on the ones that will create the most impact. It was developed by the Italian economist Vilfredo Pareto. By applying this rule, you concentrate only on two out of ten of your more critical issues at hand, to solve most of your problems with the least effort.

From "Me" to "We"

Does your company have a *me* or *we* culture? Listen to yourself. When you talk about it, do you say "me" and "they" (for example, "they gave me a raise" or "we finished the project")?

During childhood, we are encouraged to share our toys, be nice to people we don't know, and make friends. How much of that do we apply to work in our adult life? Do we think as individuals or as teams?

New technologies demand continuous innovation. The number of new products makes customers pickier than ever. They want products that are customized for them. More technologies and more products bring more information and complexity, requiring multidisciplinary and global teams to handle them. By eliminating barriers and silo mindsets, we all can work together to accomplish better results.

We are living in the era of the sharing economy. Business models that thrive are those that offer services that are shared either free or for a low fee and in turn receive payment not from the individual consumer but from other private companies, thus creating markets that are open to more offerors worldwide. Google, Facebook, Airbnb, YouTube, and TikTok are great examples. For instance, hundreds of YouTubers, usually teenagers, are amassing millions by sharing their day-to-day activities online on YouTube. The total value of the cryptocurrency market, an unregulated currency that wants to put the power and responsibility in the currency holders' hands instead of bankers, hit its all-time high of \$3 trillion in November 2021.[13] In this context, organizations cannot work the same way; they need to evolve to work internally with a sharing mindset too. That's what the new generations—native creators—expect and demand.

Workplaces need to evolve from a *me* to a *we* mindset to survive. We culture companies promote a team-oriented culture that encourages employees to share more resources instead of competing and collaborating. This sense of collaboration reduces the stress employees usually feel at work, reducing burnout and increasing engagement and satisfaction.

Employees who feel better at work are able to build better products and services for their customers. Customers have a more inspiring buying experience, so they buy more often, recommend the products to others, and even contribute with ideas for future products. More sales, more innovation, and less wasted resources end up increasing the business's bottom line organically through a virtuous cycle that continues reinforcing better results over time, as shown in Figure I.1.

This approach focuses on creating value with all stakeholders, expanding beyond shareholders to include customers, employees, partners, and society in a system where everyone can win.

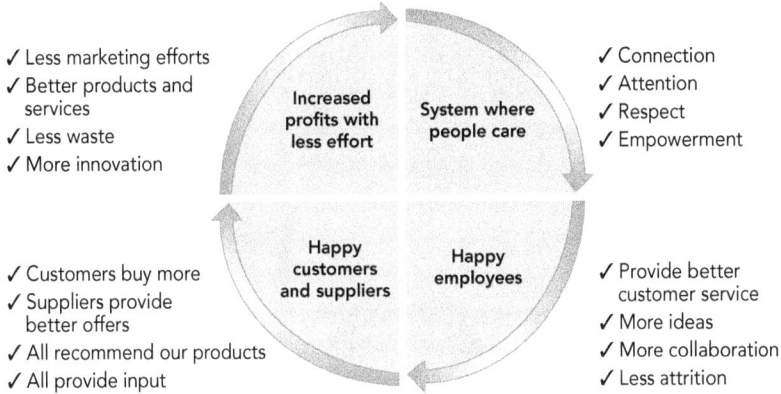

✓ Less marketing efforts
✓ Better products and services
✓ Less waste
✓ More innovation

Increased profits with less effort

System where people care

✓ Connection
✓ Attention
✓ Respect
✓ Empowerment

✓ Customers buy more
✓ Suppliers provide better offers
✓ All recommend our products
✓ All provide input

Happy customers and suppliers

Happy employees

✓ Provide better customer service
✓ More ideas
✓ More collaboration
✓ Less attrition

Figure I.1   Virtuous cycle and attributes.

CARE—Employee Engagement Model

After working with hundreds of teams and interviewing multiple company leaders, I have found 12 behaviors for the future of work to lead, care, and collaborate in a face-to-face, remote, or hybrid workplace instead of driving individualism, competition, and power. These skills are the fundamentals of the four dimensions of the CARE model—connection, attention, respect, and empowerment—to drive a We Culture. While this book does not intend to promote "one best way," these four dimensions have been identified consistently in company cultures that are thriving in this knowledge era, especially post-pandemic.

These four dimensions are intertwined and include skills from the four types of intelligence that team members need to develop through practicing the 12 behaviors that can become new soft skills if developed through training and used every day (Figure I.2).

- Intuitive intelligence to **connect** the dots to drive innovation
- Mental intelligence to **attend** the details to improve quality

- Emotional intelligence to **respect** everyone and everything to improve engagement

- Physical intelligence to walk the talk and **empower** people to act in a self-organized way to improve agility

This set of skills will increase a sense of care for the company, the team members, the community, and other interested parties.

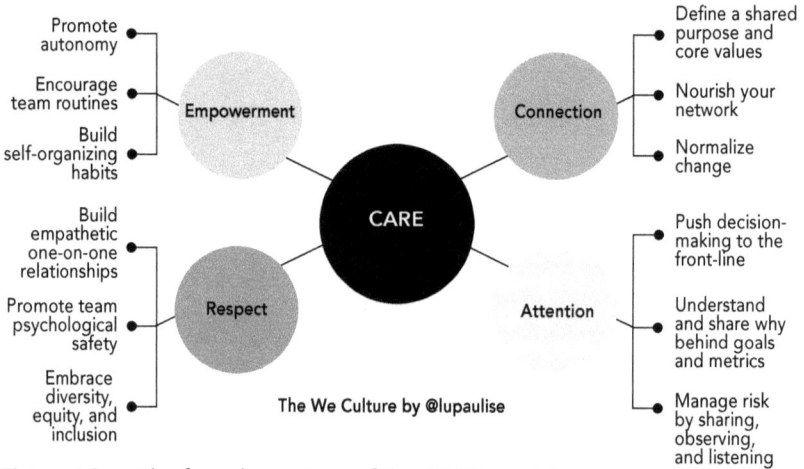

Figure I.2   The four dimensions of the CARE model.

As you have seen in the previous metrics, employee engagement drives innovation, and innovation drives engagement. *Empowerment* also drives engagement and innovation. *Attention* increases innovation and engagement, and so on. It is an integrated system to help you achieve your dreams' culture to become more productive, happier, and sustainable in the long term.

Connection, attention, respect, and empowerment are the inputs needed to move the main process of a company from a *Me* Culture to a *We* Culture: vision setting, decision-making, communication, and time management, in order to achieve innovation, quality, engagement, and agility (Figure I.3).

	ME		**WE**
Vision	Focus on short-term objectives; long-term objectives are not shared.	**Connection**	The purpose, values, and long-term objectives of the company are well-known and shared to everyone. Ideas are encouraged. **+ innovation**
Decision-making	Decision making is done at the top, informed by judgment or based solely on KPIs. Results are rewarded extrinsically.	**Attention**	Employees are encouraged to follow up metrics and analytics, understand results, and make decisions in the workshop. Intrinsic rewards drive employees even with no controls. **+ quality**
Communication	Employees are evaluated yearly, awarded based on individual performance, and competition and rankings are encouraged.	**Respect**	Multi-way collaboration, knowledge sharing, communication, and team work are promoted. Diversity and participation is encouraged. **+ engagement**
Time management	Employees check with management when there is a change in tasks, priorities, or issues.	**Empowerment**	Roles and routines are defined within the team to increase autonomy and self-organization. **+ agility**

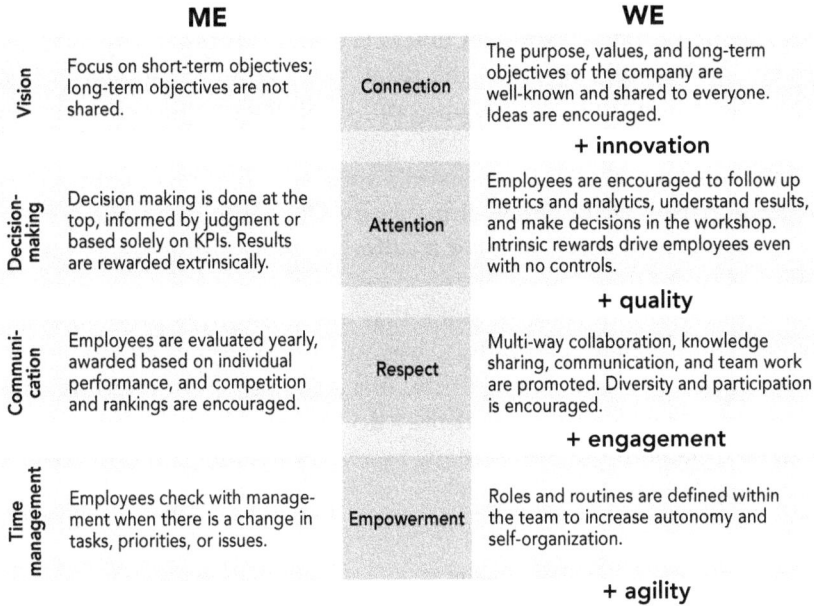

Figure I.3   From Me to We.

Throughout the book, you will learn the following about the CARE model:

1. What: The 12 best practices or skills to develop
2. How to develop them: Through 7Rs, seven elements build the context that drives your team members' behaviors (part 1)
3. How much: Measure and sense performance (part 2)
4. Where you develop them: Across the entire employee experience (part 3)
5. When: Manage time effectively to be successful long term (part 4)

Different organizations (companies, families, schools, governments, or not-for-profit organizations) have different approaches to culture. Even when they should work on these four dimensions of the CARE model—connection, attention, respect, and empowerment—each organization will have a preference based on the industry and its purpose.

Some will look to be more innovative (focus on connection); others will focus on being better at employee and customer engagement (focus on respect). It works the same way with individuals. While many recruiters and leaders may still think the attention dimension dominated by IQ is the most important to develop, IQ only contributes about 20% to a person's success in life,[14] while the rest is a combination of the other dimensions. Depending on your type of job, each dimension will have a different contribution, but none of them is better than the other.

We have all developed skills that make us thrive more in some aspects than in others. It is why teams and intercompany collaboration are important in a We Culture. Because other individuals and companies can focus on what they do best and look for others that can complement them to achieve better results faster and in a more sustainable way.

Hands-on

Exercise

Think of "me" items in your workplace and think of how you can turn them into "we" using the four CARE dimensions.

Me		We
Me		**We**
Individual office	➔	Shared open desks
_____	➔	_____
_____	➔	_____

References

4. *Google 10-K Q4 2020*, U.S. Securities and Exchange Commission, 2020, https://abc.xyz/investor/static/pdf/20210203_alphabet_10K.pdf.

5. Ben Wigert and Jim Harter, "Re-Engineering Performance Management," *Gallup*, https://www.gallup.com/workplace/238064/re-engineering-performance-management.aspx?thank-you-report-form=1.

6. Ben Wigert and Jim Harter, "Re-Engineering Performance Management," *Gallup*, https://www.gallup.com/workplace/238064/re-engineering-performance-management.aspx?thank-you-report-form=1.

7. Isolved, "Transforming Employee Experience. A SWOT analysis of 500 Human Resources Departments," *Isolved*, https://www.isolvedhcm.com/transforming-employee-experience-report.

8. James K. Harter, Frank L. Schmidt, Sangeeta Agrawal, Anthony Blue, Stephanie K. Plowman, Patrick Josh, and Jim Asplund, "The Relationship Between Engagement at Work and Organizational Outcomes," *2020 Q12® Meta-Analysis*, Gallup.

9. BCG, "How to Drive a Digital Transformation: Culture is Key," Boston Consulting Group, 2022, accessed February 16, 2022, https://www.bcg.com/capabilities/digital-technology-data/digital-transformation/how-to-drive-digital-culture, accessed February 16, 2022.

10. Tera Allas and Bill Schaninger, McKinsey, "The Boss Factor: Making the World a Better Place through Workplace Relationships," *McKinsey Quarterly*, September 22, 2020, https://www.mckinsey.com/business-functions/organization/our-insights/the-boss-factor-making-the-world-a-better-place-through-workplace-relationships?cid=other-eml-shl-mip-mck&hlkid=ee1afa0349f241cf98cbe71ab9baa938&hctky=10439215&hdpid=0ffa4915-1b11-41eb-a81f-8fb49d07bbc6#.

11. Dori Meinert, "Leadership Development Spending is Up," *HR Magazine*, SHRM, 2014, https://www.shrm.org/hr-today/news/hr-magazine/pages/0814-execbrief.aspx.

12. Mary Abbajay, "What to Do When You Have a Bad Boss," *Harvard Business Review*, September 7, 2018, https://hbr.org/2018/09/what-to-do-when-you-have-a-bad-boss.

13. "State of Blockchain Q3'21 Report," *Cbinsights*, 2021, https://www.cbinsights.com/research/report/blockchain-trends-q3-2021/?utm_source=Iterable&utm_medium=email&utm_campaign=campaign_3175269.

14. Daniel Goleman, *Emotional Intelligence, Why It Can Matter More than IQ* (New York: Bantam Books, 1995).

Part 1:
CONNECTION

We are making the culture come alive on a regular basis, in a real way. You have to walk your talk. To represent the value of SentinelOne, we do team-oriented activities. Even when you are new you get drawn into it. You are creating an ongoing experience when you actually live it. Talk is not enough; what matters is that you live these value-creating experiences that reflect those values, and reward people who are role modes for those values.

With this rapid expansion, it is important to articulate the values that define our culture: accountability (taking ownership for your actions), trust, OneSentinel (one team working together toward a common vision and purpose), community and integrity. Accountability is a big thread that unites various projects. We offer coaching through our leaders that set the example. They play a very supportive role.

2020 was a challenging year. One of our core values is community: we ran a campaign around the world to say thank you to first responders. Now we are doing one around hunger in developing countries. We not only take care of our people but also give back. It builds a good sense of community when you give back, and you build the community inside.

–Divya Ghatak
SentinelOne's Chief People Officer[15]

C onnection is the first dimension that all members of an organization need to cultivate. This is the dimension of intuitive intelligence where you master the skills to drive innovation.

The connection set of skills is key to developing the organization's "intuitive" side by linking people and ideas. The purpose and values of the organization are essential, because they are the glue that holds things together and redirects the actions. When change is constant, a reminder of "why we are together in this" helps manage conflict and stay afloat.

A lack of social connection, whether with friends, family members, or coworkers, also considered loneliness, can have serious consequences. Research indicates being socially disconnected can contribute to poor work performance, reduced creativity, and flawed decision-making and may be associated with health problems,[16] including heart disease, dementia, and cancer.

Connection During Crisis

Under typical circumstances, connection with other human beings reduces stress, increases the sense of purpose, and drives innovation. When a team works in a hybrid workplace, the impact on this connection is huge, as team members cannot see each other every day, unless they intend to. Some employees will foster the interaction even remotely, but others may be more reluctant. Teams have to generate alternatives to facilitate connection and be able to involve everyone.

Working at the same location makes it easier to "see" the connection and relationships with others. When working remotely, away from your team members, the connection is even more critical to make everyone feel safe, but it is also more challenging. Working apart, employees demand different types of communication to achieve the same results, which requires being more intentional about how to get connected.

The team needs to clarify its purpose, find ways to communicate with each other regularly and consistently, and define new ways to foster innovation and change.

The three connection behaviors that will drive innovation are:

- Share the purpose and core values
- Nourish your network
- Normalize change

The following figure (Figure P1.1) will be shown at the beginning of every part of this book to showcase the input—in this case, connection—needed to practice the behaviors that, when they become skills that everyone in the organization masters, will drive innovation as the output.

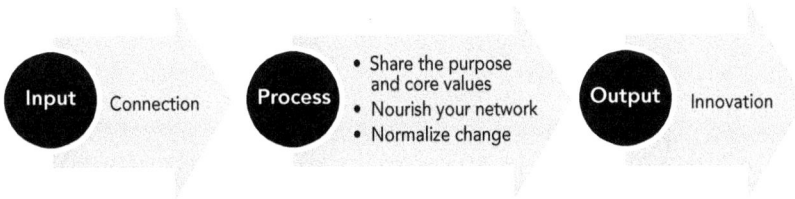

Figure P1.1 The process of the connection dimension.

References

15. Luciana Paulise, "6 Ways to Build a Culture in a Hypergrowth Company," *Forbes*, January 11, 2021, www.forbes.com/sites/lucianapaulise/2021/01/11/6-ways-to-build-culture-in-a-hypergrowth-company/?sh=7db9a3503143.

16. Constance N. Hadley and Mark Mortensen, "Are Your Team Members Lonely?" *MIT Sloan Management Review*, Winter 2021, https://sloanreview.mit.edu/article/are-your-team-members-lonely/#ref2.

1

Skill #1: Define a Shared Purpose and Core Values

Dr. Reddy's Laboratories, a 33-year-old global pharmaceutical company headquartered in India, produces affordable generic medication. With the company's more than seven distinct business units operating in 27 countries and more than 20,000 employees, decision-making had grown more convoluted, and the organization's branches had become misaligned. Over the years, Dr. Reddy's had built in lots of procedures and for many good reasons. But those procedures had also slowed the company down.

Prasad, the company's CEO, sought to evolve Dr. Reddy's culture to be nimble, innovative, and patient-centered. His leadership team began with a search for purpose. They worked to learn about the needs of everyone, from shop-floor workers to scientists, external partners, and investors. They came up with: "Good health can't wait."

They didn't just post it on motivational posters. The leadership team began by quietly using it to start guiding their own decisions. The goal was to demonstrate it. They built two innovation studios in India. After the idea of "good health can't wait" was introduced, one of the scientists developed a product in 15 days.[17]

P eople want to have a sense of purpose. Having a purpose means pursuing a mission or ideal that's larger than you as an individual; it could be for a company, community, or individual. Some people think of this as their legacy.

It is even more relevant during a crisis when everything can feel out of control. It is what helps bridge the gaps of miscommunication and lack of connection, and serves as a guiding light. Finding purpose helps put events in perspective and refocus energy on the action. Every team member needs to understand the "why" behind their job, how it could change, and how it is still relevant.

For instance, after a loved one's death, many people choose to join a cause or make donations. Likewise, a company can bring people together by supporting a cause or proposing a shared mission that solidifies the human connection beyond the crisis.

Me Practice: "Just" Posting the Values on the Wall

While there is nothing wrong with posting the values, purpose, or mission on a wall, that's not enough. Organizations have been writing and rewriting their mission and vision statements for decades. Unfortunately, these are usually just empty words. Sometimes they are only shared with the employees during onboarding training or are hanging on the manager's wall. But mission and vision alone are not enough. Researchers have shown that *purpose* is also important to guide a company. It has become a foundational element of a company's culture, the reason for team members to stay engaged. Teams need to know what needs to be done, but they also need to know why. That is the purpose — why we do it, our aim. Every system, team, or company "is a network of interdependent components that work together to try to accomplish an aim. Without an aim, there is no system,"[18] according to W. Edwards Deming.

- -

[Every system, a team or a company] is a network of interdependent components that work together to try to accomplish an aim. A system must have an aim. Without an aim, there is no system.

–W. Edwards Deming[19]

- -

Define, then, the vision of the company, which is an ambitious image of where you want to be in 20 years, and think about why you want to be there.

The Leader's Purpose

The We Culture mindset looks forward to developing all members of the team equally. Some teams may have leaders, while others may be self-organized. When there is actually a leader, their purpose is to facilitate interactions to become facilitators of success. Leaders don't need to look like heroes who know everything; they need to set the rules so everyone can become a hero at any time.

Key Insight

Leaders don't need to be the heroes; they need to set the rules so everyone can become a hero at any time.

Hence, agile leaders need to let go of micromanaging day-to-day activities.

Imagine you have to lead the organization of an event. You find yourself contacting the sponsors, contacting the speakers, defining the platform for the conference, sending the meeting invites and the agenda, and updating the minutes. Yes, you can do it all; you may even do it better than others. But if you manage all those details, who will oversee the whole conference? Who will make sure everything fits together? That's the leader's job. If you go into the details, you can't see the big picture, and nobody else will. While at the same time, you may, for instance, follow up poorly on the sponsorships due to lack of time. So, take the time to let your team help you. Clarify what they need to do, give them the resources they need or the means to ask for help while you do your stuff, and organize follow-up meetings to review the status of their tasks.

Team's Purpose

Each individual, no matter their position, age, or background, needs to understand how to connect their actions to the purpose of the team (see Figure C1.1). Let's say the goal is to give customers a unique experience. All team members need to feel they are contributing and ensure their roles and responsibilities are clear.

Key Insight

Team members need to fulfill their responsibilities for the system to work.

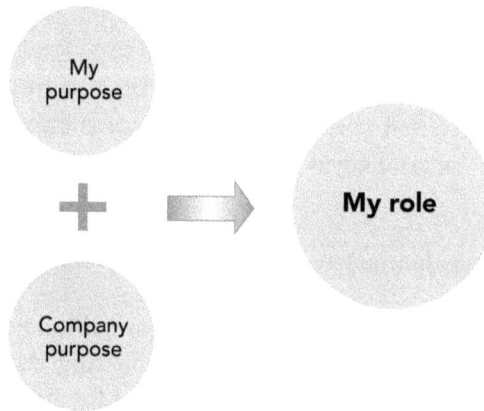

Figure C1.1   The purposeful connection.

Employee's Purpose

Your individual purpose also needs to align with the team and the organizational purpose. Whatever your position, it is important to take a step back and reconnect with yourself. What are your values? What are your strengths? What drives you and why? What skills do you have, and what is the gap with the required skills?

Key Insight

Team members need to understand how their actions connect to the company purpose and their own purpose.

Why a Company Purpose Can Become a Competitive Advantage

If you ask any company to state its purpose, the usual response is "to make money." Of course, employees are also at work to make money, but that doesn't make them come to work happy and engaged. That doesn't make them choose the company they want to work for. Younger generations expect something different.

Tony Hsieh said, "If happiness is everyone's ultimate goal, wouldn't it be great if we could change the world and get everyone and every business thinking in that context?"[20] The company he founded, Zappos, is one of the most recognized companies gaining a competitive advantage in the market by putting culture first and focusing clearly on *delivering happiness to the world*. That is its purpose, and its customers know it.

Another example, the car manufacturer Toyota, says its purpose is to "create long-term mutual prosperity for all stakeholders." Wouldn't it be great if all the companies in the world could work to help employees, customers, suppliers, and the community maximize it?

One-third of global employees strongly agree with the statement: "The mission and purpose of my organization makes me feel my job is important." By moving that ratio to 8 in 10 employees, business units have realized a 51% reduction in absenteeism, a 64% drop in safety incidents, and a 29% improvement in quality.[21]

The manager alone accounts for 70% of the variance in team engagement based on the

- manager's innate tendencies,
- manager's engagement, and
- employee's perception of the manager's behaviors.

Surviving Thanks to Your Purpose

Words and statements are great, but the problems, challenges, losses, and grief hit. Sometimes we lose our north when crisis strikes. We don't know what really matters anymore.

I recently suffered the loss of my dear brother. I felt vulnerable, disconnected from everything else, even from myself. Empty. After going on a two-day solo trip, I realized I could lose everything, but

I could never lose my purpose. If I would stick to it, I would never feel empty again. So I realized how focusing more than ever on my purpose could help me survive.

The same happens with teams and companies. As the research from McKinsey & Company's *Leading agile transformation*,[22] states: "Purpose amounts to a clear, shared, and compelling aspiration: the north star of the organization."

Why is Purpose So Important?

Having a higher purpose in mind makes people work harder. For example, the purpose may be to have the best customer service or the best product on the market, or it may be a more social purpose, such as being a green company.

A clear purpose allows people to:

- **Survive:** Teams and individuals are focused and optimistic about the organization's capacity to survive in the future. Team members learn to focus on what must be achieved, not on what goes wrong. What goes wrong may generate delays, but the final objective remains clear. Without a clear purpose, errors and delays turn into fear and distrust. With a clear purpose in mind, failing just becomes part of the process. At Toyota they call this the light of sight: "While there may be mitigating conditions such as economic fluctuations, material disruptions, etc., the True North, that is, the vision, purpose, and mission, always remains the same."[23]

- **Empower:** A clear purpose holds people accountable and makes it easier to prioritize. People understand how a metric impacts the purpose and why they are doing that job. The team members' goal should not be just to do their jobs but also to improve their jobs every day. A clear purpose helps them see that.

- **Connect people and processes:** Gene Krantz, flight director of the Apollo 13,[24] says in his book, "A clear goal, a powerful mandate, and a unified team allowed the US to move from a distant second in space into a preeminent position."[25] When employees know the purpose behind the work and how it is connected to the big picture, they work better together. It is not just their job that needs to be done — their piece feeds other parts of the process.

- **Engage:** In Toyota, for example, employees leave their initials printed on every part they make, increasing their sense of contribution. For some people, being part of a green company and participating in social responsibility activities, for example, helps them feel more emotionally engaged and makes them less prone to leave the organization for other reasons.

Key Insight

During a crisis, the purpose is more important than ever. It is the glue that helps bridge the gaps.

Team Purpose and Values Are At the Core of Building Team Alignment

A leader recently asked me how he could make a group of people work as a team when each of them had different points of view, values, and objectives based on their function. How could he build team alignment to accomplish the team objective?

The first thing leaders need to agree on is the team's purpose. A common purpose drives everything the organization does and guides decision-making. When there is no common purpose, each team member decides according to their own standards, which, of course, aren't always aligned. People think defining values and purpose is not necessary, as it is common sense. But common sense seems to be the least common of all the senses.

There is also much confusion among leaders concerning purpose, mission, and vision. Are they the same? Do we need all of them?

Mission and vision are created as part of the company's strategy to set a direction. The vision is the destination, or the *where*; the mission is the *what* we are doing to get there; the values are the *how*, and the purpose is the *why*. It is the why that provides the fuel to run the operations every day and overcome all the barriers (see Figure C1.2).

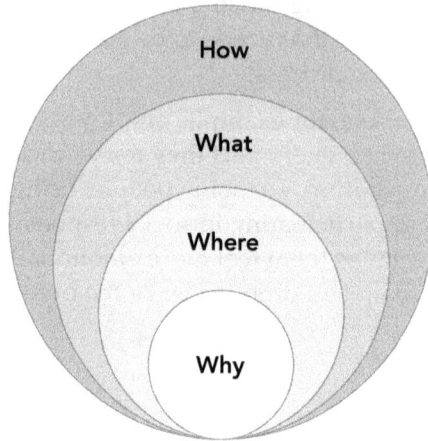

Figure C1.2   The why, where, what, and how of a company.

It is not the same to say:

- I am writing code—the person knows what needs to be done, but may not feel motivated about it every day.
- I am writing code to build an app that will help people learn leadership skills—when you know why you are coding and the impact it will have on other people, it is easier to get motivated every day to move forward.

How to Write a Purpose

Involve team members in defining the purpose. The purpose will help your team help you; hence, the owner, the executive team, or the employees can define it. Delegating to HR is less time-consuming, but it is not as effective in long-term engagement as facilitating a discussion with the entire company.

> *There can be no profit without*
> *a well-defined purpose.*
>
> –Richard Branson[26]

In the case of Dr. Reddy's, it was born thanks to a shared effort in which employees, suppliers, and customers participated. When defining the purpose as a shared effort takes longer, it becomes more meaningful to everyone involved; everyone already feels a part of it, and it's easier to remember over time. Surveys, focus groups, and one-on-one meetings can help determine a purpose with which everyone feels connected. Many companies say they discussed the purpose with their employees, but McKinsey's research[27] shows that while 72% of top leaders said they involved employees in developing the organization's purpose, only 56% of front-line employees agree. This disconnection is what drives disengagement, so make sure employees are genuinely part of the discussion.

The purpose discussion needs to start with the executive team. Most leaders aspire to a more significant social impact and want to change the company's purpose—but they don't know where to start. A coach or facilitator can help drive the discussion to be fair and impartial and help get the best out of each idea. Then, they can discuss it with the team and make sure everyone can add value while understanding how they connect their actions to that purpose.

Let's say the goal is to give customers a unique experience. First, ask your team how they feel contributing to this goal and ensure their roles and responsibilities are clear.

Like in the example of Dr. Reddy, once the company developed the idea of "good health can't wait," it was easy for the scientists to understand that the solution had to be developed quickly, which led them to develop a product in 15 days, something unheard of in the pharmaceutical industry.

Something similar happened with the COVID-19 vaccine. It would have taken more than five years to develop a vaccine if the laboratories did not have a clear *why*: The population is dying, and people can't go to work or see family members anymore. We need a vaccine at all costs.

Knowing the why is central to driving employees' efforts. So if employees are not part of developing the purpose, then communicating the purpose continuously is crucial for them to internalize it. Develop rules to internalize the purpose to guide day-to-day decisions.

Now, agreeing on the purpose is still not enough. It has to be internalized and practiced every day. Once the team organization has defined the why, it is essential to determine the how. The how is made of rules or values that help people understand the purpose, know how they need to work every day, and guide them in decision-making.

Rules are beneficial when companies grow too fast. They must be an essential part of the interview and onboarding processes to help align employees and guide them through the entire employee experience, including every meeting and project.

Here are our guidelines. As long as you operate within them, you can go crazy; do whatever you believe is best for our players.

– Brandon Hsiung, Riot Games product lead[28]

The same McKinsey research showed that an aerospace company conducted a companywide exercise to help everyone "see more clearly the behaviors they needed to change as part of a purpose initiative." The objective was to shift the focus of the company's purpose from "we fly planes" to a more customer-service-oriented mission of "we fly people." For instance, engineers saw how they were more focused on technical considerations in their designs than passenger comfort. Leaders saw how they could contribute better by shifting from a top-down command-and-control approach to a bottom-up listen actively approach.

Many companies have developed their list of rules. The sweet spot is between 5 and 12 rules, no more than that, to guide the expected behaviors and help employees make decisions daily and make them easy to remember.

Google has its eight rules (see Chapter 5). Zappos has its 10 rules. Zappos developed its rules through a common effort guided by its CEO Tony Hsieh to help direct employee behaviors. He defined the values based on inputs from all the employees. Zappos' 10 commandments[29] influence everything from hiring to merit raises and firing.

1. Deliver WOW through service

2. Embrace and drive change

3. Create fun and a little weirdness

4. Be adventurous, creative, and open-minded

5. Pursue growth and learning

6. Build open and honest relationships with communication

7. Build a positive team and family spirit

8. Do more with less

9. Be passionate and determined

10. Be humble

How Do You Internalize the Rules?

Zappos introduces its 10 rules saying: "By reading, studying, and understanding this Oath of Employment, you agree to enact each item and agree to embrace each item in energizing your roles and in your decision-making."

Each rule requires you to ask yourself what it means to you and offer three sample behaviors for the value.

Example for the value #1: Deliver WOW through service.

Ask yourself: What things can you improve upon in your work or attitude to WOW more people? Have you WOWed at least one person today?

Sample Behaviors For This Value:

- Help even when it "isn't your job."
- Wow everyone, everywhere.
- Actually make me say, "Wow."

Prioritize the Rules and Provide Examples to Avoid Inconsistencies

The values need to be clear, prioritized, and ranked based on the order of importance and have defined behaviors to bring consistency. Companies can prototype the values or provide a day-to-day toolkit with employees' feedback, actions, and stories on how they understand those values. They can even develop front-line "purpose ambassadors" to help in this process.

For example, if a value is safety, the behavior could be "I take action always to put safety first." A toolkit could include more specific actions such as "Identify, correct, and immediately report safety concerns, avoid shortcuts, and always ask 'Is there a safer way?'"

Can the Values Be Updated?

Yes, definitely, values can be updated, added, or removed based on the company's evolution. However, the changes shouldn't be taken lightly and should represent a long-term change in the company's view.

A good example is Amazon. In the early days, Amazon developed some leadership principles, which it updated over time. Before Jeff Bezos, Amazon's founder, left the company as CEO in July 2021, the company communicated two new leadership principles.[30] Since the early days, Amazon has committed to various principles: "Our Leadership Principles describe how Amazon does business, how leaders lead, and how we keep the customer at the center of our decisions."

Strive to Be the Earth's Best Employer

The first principle that was added: *Strive to be Earth's Best Employer.*

Leaders work every day to create a safer, more productive, higher performing, more diverse, and more just work environment. They lead with empathy, have fun at work, and make it easy for others to have fun. Leaders ask themselves: Are my fellow employees growing? Are they empowered? Are they ready for what's next? Leaders have a vision for and commitment to their employees' personal success, whether that be at Amazon or elsewhere.

To actively respond to the recent challenges with unions in Alabama's warehouses, the letter stated, "It's clear to me that we need a better vision for how we create value for employees—a vision for their success." The company will also increase safety, focusing, for example.

Success and Scale Bring Broad Responsibility

The second principle that was added: *Success and Scale Bring Broad Responsibility.*

We started in a garage, but we're not there anymore. We are big, we impact the world, and we are far from perfect. We must be humble and thoughtful about even the secondary effects of our actions. Our local communities, planet, and future generations need us to be better every day. We must begin each day with a determination to make better, do better, and be better for our customers, our employees, our partners, and the world at large. And we must end every day knowing we can do even more tomorrow. Leaders create more than they consume and always leave things better than how they found them.

Bezos mentioned in the letter that "Smart action on climate change will not only stop bad things from happening, it will also make our economy more efficient, help drive technological change, and reduce risks." He said that the coming decade would be decisive, and Amazon would be at the heart of the change, making progress toward the goal of 100% renewable energy by 2025.

Key Insight

Define no more than 12 ground rules, and *never* break them.

We Culture Tool: The 7Rs

To bring alignment across teams, the company needs to agree on the main rules of the road that leaders and team members will follow to model the culture. The rules of the road need to be defined at two different levels: the entire company and the specific project. Of course, the purpose of the project should be aligned to the company's purpose.

The We Culture will not be the result of a cut-and-paste exercise. Every company develops its own culture with its own specific set of rules. The company's We Culture will be customized based on the input from the 7Rs (see Figure C1.3).

Figure C1.3   The 7Rs.

Reason: The purpose is the reason why employees will work for the company and the specific project. Employees no longer take tasks as they are. To engage with the team, they need to understand the reason behind it.

Rules: Teams need to be disciplined to be able to accomplish. They must have unambiguous rules or a shared set of values to make decisions accordingly and help team members have autonomy in their jobs. Culture and values are also useful to hire and predict employee disengagement.

Routines: Routines are sequences of actions followed regularly. Routines help people execute actions consistently the same way to reduce mental effort. Over time, people perform the actions by heart, unconsciously. Every company and every individual has routines, but not all of them are productive or aligned to the purpose. Setting intentionally certain everyday routines is crucial to developing the 12 CARE behaviors or other positive and desired behaviors. Which routines should be kept and which ones should be avoided or changed? For example, establishing 15-minute daily meetings to share ideas, communicating instantly when something is wrong, or cleaning a room when finishing a meeting are the types of routines that should be set intentionally and promoted to everyone in the organization. Every daily practice is essential to translate the values into tangible actions.

Roles: The organization will need to define whether the different employee roles or expected functions will be pre-established for specific positions or determined by each project. How much power are leaders going to have to make decisions?

Rewards: The we culture paradigm promotes a win-win approach to rewards and negotiations. Sharing results in a transparent way helps employees check how well they are accomplishing the purpose. The way the company provides those rewards will drive team or individual behaviors. Are targets going to be predefined by management or set by employees? And what about the rewards? Who will define how to distribute them if there are any?

Risk-taking approach: The risk-taking style is key to determining the level of innovation the company will have. The aversion to risk will vary in a technology company versus a manufacturing company, even if both companies want to have a We Culture. Traditional teams usually avoid risk-taking by moving decision-making to the upper levels. On the contrary, the We Culture companies promote pushing decision-making to whomever is closest to the problem. "We" are all capable of making decisions in our areas of expertise, but innovative companies will usually take more risks in the pursuit of innovation than manufacturing or healthcare organizations, such as developing and testing products that may not work, but will also take actions to mitigate those risks.

Reinforcement: The last ingredient is how these routines will be reinforced. How are companies going to make sure they repeat, repeat, and repeat the expected action until it's natural and unconscious? One person requires time to stick to a new habit, and it is even harder for various teams. But the good news is that habits are contagious, so disciplined people repeating the same behaviors can build momentum faster than only one. If we focus less on what team members are doing wrong or what's missing, and more on what is available and its potential, we realize that everyone will be able to help the company purpose in their own way. We shift from a deficit to a strength-based paradigm, where stress unleashes innovation.

Hands-on 1.1

Meet with your team members and define your team's purpose. Then define the ground rules that will help you "stick" the group together. Set no more than 12 ground rules, post them where everyone can see, preferably online and onsite, and define each individual's responsibility to accomplish that purpose.

Exercise

Defining ground rules that help STICK the group together.

Define a purpose with your team.	Set no more than 12 ground rules.	Define your responsibilities.
_____	_____	_____
_____	_____	_____
_____	_____	_____
_____	_____	_____

References

17. Bryan Walker and Sarah A. Soule, "Changing Company Culture Requires a Movement, Not a Mandate," *Harvard Business Review*, 2017. https://hbr.org/2017/06/changing-company-culture-requires-a-movement-not-a-mandate.

18. W. Edwards Deming, *The New Economics: For Industry, Government, Education*, second edition (Cambridge, MA: The MIT Press, 1994), 50.

19. W. Edwards Deming, *The New Economics: For Industry, Government, Education*, second edition (Cambridge, MA: The MIT Press, 1994), 50.

20. Tony Hsieh, *Delivering Happiness* (New York: Grand Central Publishing, 2013).

21. Gallup, "Gallup's Perspective on Designing Your Organization's Employee Experience," *Gallup*, 2018, https://www.gallup.com/workplace/355601/employee-experience-paper.aspx.

22. Aaron de Smet, Michael Lurie, Andre St George, "Leading Agile Transformation: The new capabilities leaders need to build 21-st century organizations," *McKinsey & Company*, 2018.

23. Tracey and Ernie Richardson, *The Toyota Engagement Equation,* Indian Edition (India: McGraw Hill Education, 2018).

24. Gene Krantz, *Failure Is Not an Option: Mission Control From Mercury to Apollo 13 and Beyond* (New York: Simon and Schuster Paperbacks, 2000).

25. Gene Krantz, *Failure Is Not an Option: Mission Control From Mercury to Apollo 13 and Beyond* (New York: Simon and Schuster Paperbacks, 2000).

26. Alp Mimaroglu, "What Richard Branson Learned From His 7 Biggest Failures," *Entrepreneur*, 2021, https://www.entrepreneur.com/article/295312.

27. Arne Gast, Nina Probst, and Bruce Simpson, "Purpose, Not Platitudes: A Personal Challenge for Top Executives," *McKinsey Quarterly,* December 3, 2020, https://www.mckinsey.com/business-functions/organization/our-insights/purpose-not-platitudes-a-personal-challenge-for-top-executives.

28. Darrell Rigby, Jeff Sutherland, and Andy Noble, "Agile at Scale: How to Go from a Few Teams to Hundreds," *Harvard Business Review*, May–June 2018, https://hbr.org/2018/05/agile-at-scale.

29. *Zappos Insights,* https://www.zapposinsights.com/about/core-values.

30. Amazon, "Amazon Announces Two New Leadership Principles," *About Amazon,* 2021, https://www.aboutamazon.com/news/company-news/two-new-leadership-principles.

2

Skill #2: Nourish Your Network

(Like in a garden) The outcome is less dependent on the initial planting than on consistent maintenance.

–Stephen Denning[31]

S kill #2 is to communicate regularly with the team's network to nourish it and build a stronger connection. Building a strong network today is essential for success. One person alone—or even a team—does not have all the information to solve an issue. Therefore, the ability to connect with others expands the possibilities to perform in any situation. One of the reasons why productivity decreased in many companies while working remotely is because the power to trigger ideas through informal connections in open offices and in-company cafes dissipated.

The Power of the Network

A *system* is a network of interdependent components that work together to try to accomplish the aim of a system. Every team, company, and family is a system.

For a network to be effective, everyone should be connected as part of a process: there is always a client and a supplier (see Figure C2.1). Every time you do something, there is somebody else afterward. Leaders need to help the team see the connection with the group—how they impact each other and how their output impacts other teams. If there is a defect and the team doesn't correct it, the next team will. If not, the customer will be the one detecting it, which will be more costly.

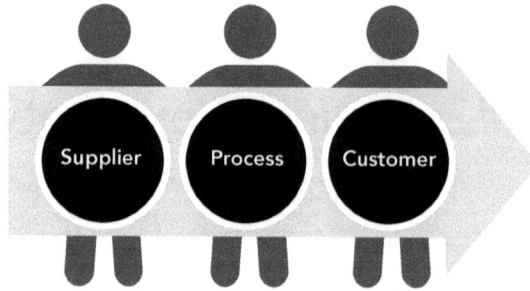

Figure C2.1   Everyone is part of a process.

When people feel like they belong to a network, they are more productive, motivated, engaged, and three and a half times more likely to contribute to their fullest potential.[32]

It takes a collaborative and inclusive approach to tackle the most challenging problems. For example, when facing a threat like a pandemic, communicating and sharing resources within the network is critical for survival.

Identify Your Stakeholders' Network to Accomplish the Vision

Everyone you know is connected somehow in this network. Who is part of your closest network? Who is not that close but is still part of your network? How do you take care of each of them? Make a list. These could be your direct reports, clients, government institutions, or co-workers. Basically, they will be the people who can help you grow continuously and help grow the business unit.

Of course, they will not all be at the same level; some will require a daily or weekly touch point, and some will require a bi-monthly call. Try to identify precisely how much time you should spend with each stakeholder, and evaluate periodically if you are doing OK or if you should be paying more attention to them.

Sometimes, we hardly talk to contacts who may be the source of potential gains and happiness in the future. On the other hand, sometimes we go to weekly meetings that are not helping us accomplish anything. Make sure you redefine your priorities and regularly keep the right contacts on the radar.

Me Culture: The Pyramid Organizational Structure

Many companies still value the pyramid format, which shows how communication should be directed. Associates expect leaders to tell them what to do. They can't communicate with the boss of their boss, because that is disrespectful. Reaching out to a customer, a supplier, or even a different department could be more than adventurous, it could be outrageous.

Some employees seem to be more important in this paradigm than others, especially further up in the pyramid. Everyone needs to listen to the most important person at the top of the hierarchy.

In his book *Reinventing Organizations*,[33] Frederick Laloux uses different colors to describe organizations. He refers to these types of companies as "orange organizations." Today, orange is the dominating view of most leaders around the world. They are project- and process-driven. Rationality is valued above all, while emotions are kept behind a mask "so that we don't make ourselves vulnerable," says Laloux. Tasks are put over people. Results, growth, and milestones are all that count. The company is like a machine.

We Culture: The Diamond Metaphor For a Successful Team

Other companies discovered the inverted pyramid, where the customer is at the top, and leaders are at the bottom as servant leaders. It would be the "green paradigm" for Laloux. Green strives for bottom-up collaboration. "Orange glorifies decisive leadership while green insists that leaders should be in service of those they lead." Examples of these companies are Southwest Airlines and Ben & Jerry's. The company is more like a family. That's why many orange leaders tend to see these companies as naive, more of a fit for nonprofits.

Think about it: Is there really someone more important than another? Some employees build the products. Others sell the products. Some need to clean the offices and stores. Leaders facilitate the efforts, and customers buy those products. Can we prescind any of them? No. Actually, they all seem to be pretty important. When analyzing your processes using Figure C2.1, you may notice that it is not easy to connect with others to accomplish your purpose when you don't know how your processes are interrelated, or how their input impacts your output and vice-versa. Again, we are all connected for a reason.

The COVID-19 pandemic has taught us that successful companies are those that are more agile to confront change, and they do so by setting a structure that pays attention to everybody. Changes come at all times and from different directions. It is not about who is on top but about how all these components interact more successfully. Laloux describes these organizations as Teal, leaving systems where change comes from everyone, everywhere, and anytime, without a central command.

As Laloux writes, "Organization's real structure is an intricate web of fluid relationships and commitments that people engage in to get the work done. Unfortunately, most organizations force a second structure, the one with boxes piled up in a pyramid shape, on top of the first."[34]

As McKinsey research[35] notes, "Linear, process-based tools such as activity-based costing, business process reengineering, and total quality management have long been effective at measuring and improving the efficiency of people and organizations in accomplishing individual tasks. But they do a poor job of shedding light on the largely invisible networks that help employees get things done. This blind spot has become problematic. Collaboration within and among organizations is more important than ever."

Many quality management practices like lean actually focus on people, but they are not always implemented emphasizing the change on people. The change is mostly focused on processes, which seem easier to measure and control. But culture is not what is written on a flowchart or a desk manual. It is what people actually do with the flowchart and the desk manual. Is your team's behavior like a pencil or a diamond?

My research shows that while organizations are like webs, the behavior of successful teams within these networks resembles the behavior of the diamond components.

Diamond is the hardest naturally occurring material known due to its strong bonds. It has a high refractive index and moderate dispersion, which gives cut diamonds their brilliance, which is the reflection of the light of each carat. All the carats of the diamond are connected and depend on each other to make the diamond strong and shiny. The carbons used for diamonds are the same ones that build the graphite that is used for pencils. The only difference is how these carbons interact with each other.[36] The higher the interaction, the stronger the material. Over time, these connections can dissolve, and the components can go somewhere else and form another diamond (or pencil).

We organizations are successful and innovative because each of their employees can shine at their highest potential. In traditional structures, only managers have the authority to make decisions. In we cultures, everyone has the authority to make changes for the company and make decisions about how they do their work. Leaders may need to "unlearn what it means to be a leader," as Stephen Denning describes in his book, *The Age of Agile*,[37] because they don't have to be heroes anymore.

Key Insight

In We Cultures, everyone has the authority to make changes to the company and make decisions about how they do their work.

If we want people to be involved and engaged, we need to shift our mindset from "How do I escalate the pyramid?" to "How do we help each other to make the diamond stronger?" We can only achieve the latter by learning to work as a team. You can imagine an organization like a diamond in Figure C2.2, but if you look at the diamond from the top, that is the client's perspective. In Figure C2.3, you should be able to see no levels of importance, just people working together for a common aim.

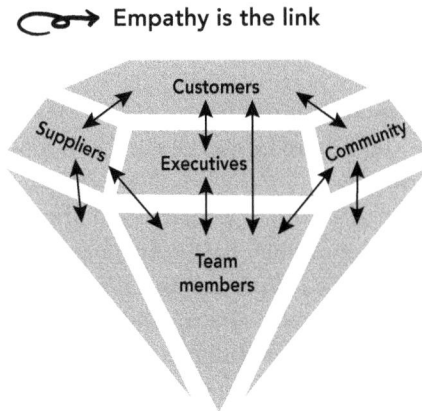

Figure **C2.2**   The Diamond methapor for a team.

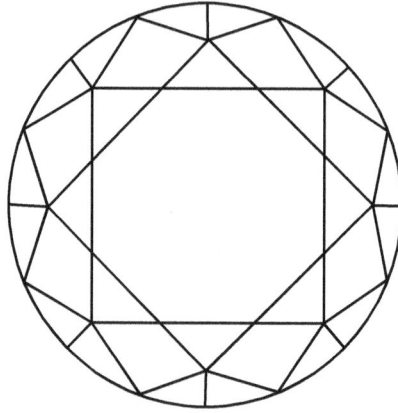

Figure C2.3   The Diamond methapor for a team, seen from the top (from the customer perspective).

Me Practice: Favor Face-To-Face Workers

Many company cultures tend to favor employees who work onsite, giving them more opportunities to grow, better salaries, or more participation.

Now that employees have tasted the sweetness of remote work for a long period, it is hard to accept an entirely onsite working model, given that they have seen how much time and resources are saved by working from home. Working from home full time also has downsides, such as increased mental health issues and reduced innovation, if not managed effectively, but it opens up the potential hiring pool to many more people. Some people with disabilities or caregivers have more flexibility and opportunities to work.

The ideal workplace model for most companies looks more like a hybrid work model where employees can choose to work both in-company and remotely.

The COVID-19 pandemic has disrupted the workplace so drastically and at such a rapid pace that going back to the pre-pandemic normal is not an option. Companies and leaders need to understand the new challenges, embrace them, and adopt new behaviors to thrive in the new normal. The skills required to lead in a face-to-face environment may not be the same in a remote or hybrid work model, so leaders need to be trained to face these challenged while avoiding favoring anyone.

There is no one-size-fits-all approach, but the new technologies allow choosing the perfect option between the face-to-face and remote continuum. While, in many cases, innovation and healthy distractions were triggered by random encounters in cafeterias and meetings, a hybrid workplace needs to find ways to connect and achieve these too.

We Culture: Facilitating Interactions

Essentially, to improve the effectiveness of the communication in our teams, leaders need to facilitate interactions top-down and bottom-up, in all directions:

- Make it easy to connect with anyone within your network. Forget about hierarchies and departments. Break down the silos by making everyone feel safe to connect with anyone they need for any project without having to contact you first.

- Once you connect with someone, build on that interaction to make it stronger. Get to know more about the person beyond their name and function.

- Use all the means possible to communicate. Some prefer calling while others prefer texting.

- Use synchronous communication (you interact with other people at the same time, such as meetings, phone calls, coffee chats, and so on), asynchronous (actions that you perform on your own and not in real-time with others — they may reply later).

- Consider time differences, language barriers, cultures, and work-life balance. There is no manual to do this, just ask the other person what is best for them.

Ning Wang, CEO at Offensive Security, said when I interviewed her, "When working in an office, almost all communications are synchronous: meetings in a conference room, one-on-ones, chit chat. When you have a distributed team or employees in 15 time zones, you cannot rely on synchronous communication only. You cannot rely on calling or slacking them. You also must design ways that promote effective asynchronous communications that are casual, fast, fun, and informative. We use email for things that are more formal. We use a team chat and collaboration platform for all frequent and quick communications, which allows people not directly engaged in the communication to digest asynchronously."

Watch Ning Wang Interview on YouTube: https://youtu.be/kUxAUauBwqw or download the We Culture app to watch other interviews.

Connection in Hybrid Teams

When individuals work remotely, they may not look forward to connecting to big masses, but they may prefer to maintain the connection with a small tribe or a team.

Improving high-quality connections[38] is essential for combatting workplace loneliness, as Jane Dutton and her colleagues at the University of Michigan's Center for Positive Organizations found out. Loneliness brings health problems, reduced productivity, turnover, and burnout, so companies sometimes need to foster more open, emotional, and personal interactions to build these high-quality relationships at work.

Following are some of the challenges in a hybrid workplace:

- Remote workers are often overlooked during meetings, especially when all the leaders are in the meeting room, onsite whiteboards and flipcharts are used, and some issues discussed in the meeting room are not heard over the microphone.

- Promotions and assignments favor onsite employees. Quora's CEO said in Quora's blog[39] that "in most companies, it is a significant career advantage to work from the headquarters rather than to work remotely. People in positions of power have a tendency to bias toward giving out opportunities to those whom they are familiar with."

- Employees working from home get demoralized when they are left out of decisions and projects and receive less feedback and positive reinforcement from their immediate leaders.

Leaders need to work on modeling certain aspects of the culture to make it more welcoming and diverse for remote, onsite, and hybrid workers. Start analyzing the support systems that help teammates work together.

The Argentinian company Globant, for instance, developed StarMeUp OS, a system that can recreate casual encounters. Employees can ask for slots to meet people with specific characteristics or interests within the company. The software will trigger random meetings with other employees worldwide in what they call "virtual chillouts." While they meet, they will also access "augmented information" on their Zoom account on who they are, what they do, or how many stars they have.

> *Culture is an imitation game. In the early days, we developed a manifesto with our core values. Still, the challenge was how to communicate it to the employees. Globant was growing too fast in many different locations. We decided to use our own technology to solve it.*
>
> *–Martin Migoya,[40] Globant's CEO*

StarMeUp OS also allows Globant to identify the cultural traits that are more successful at the company and help recruit people who show the same patterns. Likewise, it can even predict attrition by analyzing the engagement curves of people who left the company. Now the company can tell when someone is disengaged, enabling leaders to start a conversation before an employee decides to leave the company.

> *Watch Martin Migoya's interview: https://youtu.be/H8hpafCgtr or download the We Culture app to watch other interviews.*

Nourishing the Network
In Challenging Situations

The pandemic raised the bar for communicating sensitive issues. Like many others during that time, there was a company that made severe changes in its HR policies, such as eliminating its retirement match and increasing the percentage of employees who had to leave the company after a poor annual performance review. This information was leaked to the public before many employees received the news, which caused frustration and fear amid the COVID-19 pandemic crisis, especially in a company that praised its retirement benefits.

Regular communication is key to avoid FOMO (or the fear of missing out). A lack of communication breeds unease and uncertainty. As a result, trust and psychological safety can quickly erode. Communicate constantly. Even when you don't have the answers, sharing what you do know builds confidence and a sense of common experience. Feeling safe is a basic human need, and crises challenge this need, especially if there is a lack of communication.

Communicate the information you can and explain what and why you can't reveal. Consider finding the answer together.

It is in these times that companies need to communicate everything as soon and as often as possible.

1. Identify what you do know. Clarify what is known and the core assumptions at play. This will build both a collective understanding and the confidence needed to move into the unknown.

2. Make a list of things you don't know but will need to better understand. Such a list will force you to get out of your own bubble and take a critical look at what is going on around you. If you don't admit you have a problem, you'll never find a solution.

3. Engage your team in finding the right problems and opportunities.

We Culture Routines

- Use all the tools possible to communicate. The most common virtual meeting tools are Zoom, Microsoft Teams, Google

Meet, Webex, and BlueJeans to interact personally with co-workers. More than 60% of communication comes from nonverbal cues, so use video.

- Turn on the video in hybrid team meetings: Most of the time when someone joins a meeting with more than 10 people, that person tends to turn off the camera and multitask while listening. To avoid distractions, ask that all team members turn on the video and maintain eye contact as much as possible to increase engagement.

- Try to have more meetings than before, but shorter and with fewer people. When working from home, people prefer to be more connected with fewer people, forming more than a team — a tribe — that communicates often.

- Try to define team meetings to check in daily, at the same time, to maintain a routine and ensure a safe space. To increase engagement, reduce meetings to 30 minutes or less, and invite every member to speak up.

You can also organize informal meetings for larger groups to promote networking. Many companies organize events such as virtual breakfasts, lunch breaks, cooking nights, birthday celebrations, or baby showers through Zoom. Have a channel open where people can connect and see how others are doing. Take advantage of the technology and organize breakout rooms for smaller groups. These meetings have no agenda; everybody can dial-in. It's a chance for employees and the management team to connect. The key is to create different ways of integrating everybody. The associates are free to find the one that suits them the best.

- **Organize to work from home at least one day a week and continue doing virtual events.** The goal is to show other employees they don't always need to be in the office. Some leaders insist on going to the office for meetings, but they are simply showing that being remote is not enough to succeed. Attend some meetings from home, not only to show it is OK, but also to understand what the challenges of working from home represent. Asana's co-founder, for instance, said its "no meeting Wednesday" policy will evolve to include the option of "work from home Wednesday."

- **Help improve the experience of remote workers during meetings by standardizing certain behaviors.** For instance, forget about the flipchart and use online brainstorming tools. Keep using an online link for every session and demand others do so too. Make sure every conversation, task assignment, and agreement includes both onsite and remote participants. A Harvard article[41] suggests that teams may need to "reestablish the team's mission, set explicit interaction norms, consistently enforce them, create a shared team identity, make roles and processes transparent, stabilize the membership, and reduce cross-team switching costs."

- **Follow up with employees more closely to compensate for the lack of interactions and emotional cues.** One-on-one meetings may need to be at least once a week; team meetings may need to be held every day for at least 15 minutes to ensure team cohesion. Leaders need to choose when it is better to have online meetings or asynchronous communications and when face-to-face is preferred, such as when building trust or discussing long-term planning and private matters.

- **Find other means to communicate.** This can include Slack channels, where you can define a topic or a question and people can contribute, and WhatsApp to send text messages and voice messages.

- **Find serendipity time.** Companies use platforms like StarMeUp or Donut to help connect individuals and teams serendipitously for virtual coffee, peer learning, or diversity, equity, and inclusion (DEI) discussions. These systems create random meetings based on certain interests defined by the individuals, similar to a dating app but with business/relationship building purposes.

- **Consider gratitude sharing.** At the start of a meeting, you can ask each person to share something he or she is grateful for to promote openness. For example, Hilary Hendricks and fellow researchers at the University of Michigan and Notre Dame are studying the use of gratitude circles at a restaurant chain. Before the lunch shift, employees gather in a circle. One member is randomly chosen to stand in the circle to have peers describe aspects of him or her that they like and admire. Early results indicate that gratitude givers and receivers can emerge from this practice feeling more connected.[42]

Key Insight

Especially in times of crisis, people can feel detached and aimless. Reducing emotional isolation, creating common purpose, and staying connected to one another personally and professionally is critical. When people, teams, and organizations connect during crisis, they emerge even stronger.

Hands-on 2.1

Draw yourself in the middle of a sheet and then draw circles like planets that orbit around you. If they are close, you should talk to them often; if they are not close, just keep them in your loop and check in with them every month or every quarter. Then, on the first of every month, study your stakeholder's galaxy and identify areas for improvement: contacts you need to get in touch with more often or reinforce and those you need to see less often.

Exercise

Draw circles of influence and write down the names of the members in your organization that work with you and draw lines to simulate how they connect with each other. Use arrows to show if they get or receive information.

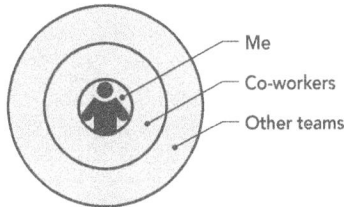

Me
Co-workers
Other teams

References

31. Stephen Denning, *The Age of Agile* (New York: Amacom, 2018).

32. Karyn Twaronite, "The Surprising Power of Simply Asking Coworkers How They're Doing," *Harvard Business Review*, 2018. https://hbr-org.cdn. ampproject.org/c/s/hbr.org/amp/2019/02/the-surprising-power-of-simply-asking-coworkers-how-theyre-doing.

33. Frederic Laloux, *Reinventing Organizations* (Brussels, Belgium: Nelson Parker, 2014).

34. Frederic Laloux, *Reinventing Organizations* (Brussels, Belgium: Nelson Parker, 2014).

35. https://www.mckinsey.com/business-functions/organization/our-insights/mapping-the-value-of-employee-collaboration.

36. *Quality Progress,* March 2020.

37. Stephen Denning, *The Age of Agile* (New York: Amacom, 2018).

38. Constance Noonan Hadley, "Employees Are Lonelier Than Ever. Here's How Employers Can Help," *Harvard Business Review*, 2021, https://hbr.org/2021/06/employees-are-lonelier-than-ever-heres-how-employers-can-help.

39. Adam D'Angelo, "Remote First at Quora," *Quora Blog*, 2020. https://quorablog.quora.com/Remote-First-at-Quora.

40. https://www.forbes.com/sites/lucianapaulise/2020/07/16/how-to-augment-collaboration-and-productivity-working-remotely/

41. Constance Noonan Hadley, "Employees Are Lonelier Than Ever. Here's How Employers Can Help," *Harvard Business Review*, 2021, https://hbr.org/2021/06/employees-are-lonelier-than-ever-heres-how-employers-can-help.

42. Constance Noonan Hadley, "Employees Are Lonelier Than Ever. Here's How Employers Can Help," *Harvard Business Review*, 2021, https://hbr.org/2021/06/employees-are-lonelier-than-ever-heres-how-employers-can-help.

3

Skill #3: Normalize Change

People actually don't resist change; people resist being changed. If you want to minimize resistance to change, involve people. Listen to them, help them to understand the need for change, help them to get involved in planning the response to the need. The more they can understand and the more they can participate, the less resistance you'll get.

– Peter Sholtes[43]

Skill #3 of the dimension of connection is to learn to normalize innovation and change. How do you drive change? How do you tackle resistance from employees for changes due to innovation? To become agile and adapt to any situation, you cannot just rely on the past; you need to connect those experiences with the current information.

Steve Jobs suggested in his 2005 Stanford commencement address[44] "connect the dots," which means observe your previous experience and use it to build your future. In order to visualize the purpose, the values, and the network, imagine something bigger than yourself, probably something impossible, and operate beyond your comfort zone. Embrace change, overcome frustrations and short-term issues, and be optimistic about a brighter future. Therefore, you may have to build new muscles, transform your culture, and embrace new methodologies to ensure the organization is resilient enough to ride the new wave of change. You can train your muscles to become more comfortable performing in new situations by doing new things, such as learning new hobbies. Make time to explore, pursue a new experience, and, if possible, put yourself in the position of student, novice, or beginner.

Keep a positive attitude to remove fear and attract more innovation. If you are in a crisis, it means a change is coming. Normalize it by accepting that crises are normal and it's not the end of the world. Crises always bring something good and bad. Accepting what you cannot control gives you more power and energy to focus on what you can control. You can control how you react, try to reduce the bad, and take more advantage of the good.

. . . .

The day before something is a breakthrough, it is a crazy idea.[45]

– Peter Diamandis,
co-founder of Singularity University

. . . .

The contact lens company 1-800 Contacts has five rules, one of which is "live for the impossible." I talked to Phil Bienert, chief marketing officer, during an interview (watch the YouTube video below). He explained that if you have a problem that anyone thinks is impossible, you need to figure out how to solve it. Before 1-800 Contacts, nobody could buy contact lenses online or take vision exams for free. So the company redesigned the prescription process and developed ExpressExam—an app that allows customers to renew their contact lens prescriptions through a 10-minute online vision exam from home.

. . . .

*Watch the interview with Phil Bienert
https://youtu.be/rpXVAPNZjVg or download
the We Culture app to learn more.*

. . . .

In the United States it is normal for doctors to send prescriptions online to pharmacies, but in other countries this is not the case yet. You still have to visit the doctor, ask them to write the prescription, and then go to the pharmacy.

Change Management Is Like Grieving

The need for change is constant. Unfortunately, when we grow old, the ability or willingness to change shrinks. The same happens in traditional companies as they grow older. Research shows that only 30% of all changes are successful,[46] because people grow weary of changing when efforts are unsuccessful or unsustainable over time. In addition, the flavor of the month's initiative frustrates employees and makes them less eager or excited to commit to change.

Some challenges employees face before the change occurs include:

- Fear
- Not feeling heard in preparation for or during the change management
- Battle fatigue, after struggling many times before during previous changes

Growing and changing is usually painful. Change is like grieving — you need to let go of something you were used to. Nobody enjoys grieving, which is why it is so difficult for teams to overcome change Unless you learn to normalize it.

The process of grieving is often described as having stages or cycles (see Figure C3.1). People may go through the stages in different sequences and go back and forth as they process loss and adjustment. The person usually passes from denial to anger, confusion, depression, and, finally, crisis before eventually starting acceptance and new confidence of what is to come.

Basically, we are constantly fighting to keep the homeostasis, that is, to keep a relatively stable balance between different independent elements in our lives the same way we do in our bodies. We have to maintain a specific temperature, heart rate, weight, and breathing rhythm to continue functioning. In our lives, we have a routine that involves our job, family demands, the environment, and so on. But things happen every day, both good and bad, that often push us away from our balance points unconsciously.

Things that happen in your life can be either expected (you have some control over the outcome) or unexpected (you cannot control the outcome).

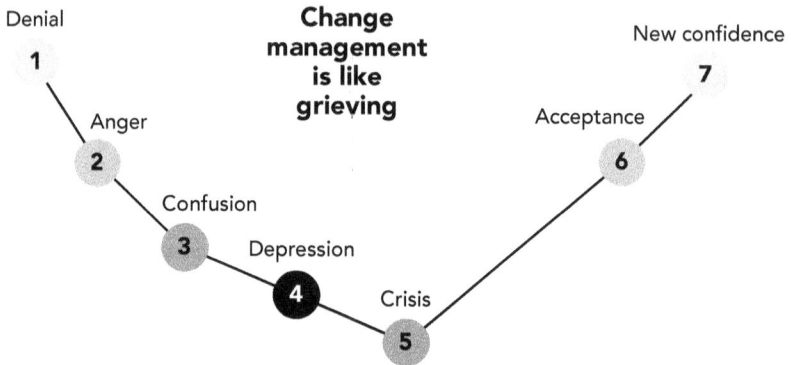

Figure C3.1 Stages of grief.

We fear the unknown. First, we deny the change exists and try to continue with our routine. But there are signs everywhere that demonstrate things are not the same, and that drives anger. We keep fighting to get back to how things were before, but we soon realize that we can't. That's when confusion starts. We don't know what to do now, because this change wasn't in our plans. We feel guilty, unable to act, and, therefore, become depressed. We fall into a black hole that paralyzes and isolates us from everyone else until we reach the peak of the crisis.

In the stage of crisis, we don't know what to do or how to get help. Maybe someone reaches out to us in a different way and helps us see the light at the end of the tunnel. Maybe so much time on our own helps us see differently. Or, maybe we come up with a new idea. Eventually, we learn to accept the change.

Acceptance is when we begin to be OK with what happened (accept what we cannot change), take responsibility for what we can do next, and move on. We gain new confidence to work toward accomplishing new tasks. Most importantly, we are willing to change our behavior in response to what is new and required in the new circumstance.

Me Culture: Blaming For Failure

Frustration, guilt, fear, and envy—unlike trust—are emotions that hold us back. We waste time regretting, doubting, and placing blame on others, which doesn't change anything.

These feelings are necessary for a time to help us review our behavior and see what needs to change, but eventually we will need to overcome them and focus on the positive. Instead of thinking about who hurt us and getting angry, we must think about what we can do differently to react to it next time. Close the loop of negativity by attaching a positive thought to every negative one. That is, every time you recall the negative, remember the positive. Do the same with projects. If a project failed, remember what you learned about it.

Many organizations have the routine of finding someone to blame for every problem. They think this will prevent similar issues in the future. But they don't realize that they are not avoiding the issue, but preventing people from disclosing the issue.

We Culture: Accepting Failure As Part of the Innovation Process

When dealing with failure and problems, the focus needs to be on fixing the problem, finding the root cause, and preventing it by addressing the root cause. Finding someone to blame and punishing that person will only prevent transparency and candid discussions.

How can you reduce negative thoughts and not fear what is to come? By trusting. You must trust that you (and the others) are doing the best you can, and trust that no one wants to hurt you. Dig deeper to learn from it, or simply let it go. Trust that this point in time will later connect with other points to make sense of what happened (again, connect the dots). Change is painful, but it is needed to guide you in the right direction. You need it in small doses, just like a push, to move to the next stage. As long as you accept it, overcome the difficulties, and remain longer in the positive thoughts, change is productive and healthy. Some levels of stress and anxiety can actually *improve* our performance. For example, many people say they perform at their best when they're under a manageable degree of deadline pressure. But, when anxiety becomes severe, our performance will almost always suffer.

As long as you keep your purpose in mind and practice how, not just to accept but how to embrace change and be happy about it, growth will come. Transforming thoughts into actions removes the feeling of powerlessness.

Accepting failure, and even better, not seeing failure as something bad but just a part of the process, helps people become more creative in the long term. If people are punished when something doesn't go as planned, they will avoid trying or proposing something new, something innovative, altogether. So failure, being wrong, and trying are critical elements in the exploration process to achieve innovation.

Key Insight
Failure, being wrong, and trying are critical elements in the exploration process to achieve innovation.

You Can Dictate Procedures, But You Cannot Dictate Innovation

While change is difficult, innovation is even more difficult. Creativity is about coming up with a big idea, while innovation is about executing the idea — converting the idea into a successful business. Innovation is applying creativity to do something profitable (you can sell it, make money, and pay the costs), feasible (you can do it), and real (people want it, there is a market).

The problem with innovation is that it cannot be dictated. You can dictate people to follow a process, but innovation is different. You cannot simply ask people to come up with the best ideas, but you *can* build and support a context where innovation can flourish organically.

Managers want good ideas to flourish, but ideas are usually stuck somewhere and never get in the right hands. Usually, employees have lots of ideas for improving their processes, using different tools, or improving customer satisfaction. The first urge is to mention them to the middle manager. What happens next?

What happens next is a common story. There are no processes to capture those ideas. Middle managers are afraid to take risks, be seen as weaker than the employee, or be too liberal. Ideas are seen as proof that something is incorrect now, so they are not welcomed. Some people think ideas are only for marketing or research and development (R&D). More urgent tasks arise, and ideas are forgotten. Sometimes employees are even punished, especially if ideas come after a mistake. Finally, employees learn that ideas are no good.

> *The story of innovation has not changed. It has always been a small team of people who have a new idea, typically not understood by people around them and their executives.*
>
> –Eric Schmidt, chairman, Google[47]

It's All About Building a Culture of Innovation

Given the difficulty of driving innovation, many companies depend on one smart guy, usually the founder, to develop all the new features in the company. Alternatively, others build it into the culture and make everyone innovate. A culture of innovation is what will drive innovation long term.

As Aristotle said, "The whole is greater than the sum of its parts." For a company to become innovative, you need to

- build a context where people can innovate;
- engage the entire team, not just one individual;
- set them free to explore new ways; and
- train them and give them the tools.

Build a Context

For instance, Google has bicycles with four seats so four team members can ride together instead of having a meeting in a boring meeting room. This bicycle helps team members think out of the box, see different things together, and practice a little team building in the meantime. Employees at Google post ideas, suggestions, and problems on shared boards and even inside restrooms so they can think, explore, and contribute at any time, with any team, based on their interests.

Engage the Entire Team

Some companies depend on innovation and make it part of their day-to-day routine. Then innovative behaviors need to be taught to all the employees. For innovation to happen more often, and especially more organically, everyone needs to know that their responsibility is to innovate. It is not only the job of the innovators or the R&D

department. This is not easy if you are used to following instructions to work. The company first needs to develop the roles and routines that align with innovation.

Team innovation can be easily started by periodically organizing brainstorming and retrospective sessions frequently that foster idea generation and sharing thoughts and issues openly. Even remotely, tools like MURAL and Poll Everywhere help organize, connect, and vote ideas.

Set Them Free to Explore New Ways

Steven Haedrich, CEO of New York Label and Box Works, a small company founded in 1878, could see that unleashing workers' potential was the way to improve and innovate. When I interviewed him in 2018,[48] he said, "Our key recommendation is to work to continuously expose the system to improve quality and promote innovation, unleashing workers' ideas. Everybody today wants 'instant pudding,' and that is not how it works." He mentioned that suppliers and clients see what they do and the value they provide, but they still do not practice the same methods because it takes a mindset change from the top.

As Haedrich says, change takes years of implementation, new habits, policies, and a clear purpose of serving the customer. "We started in 1993 with an external Deming consultant." The company decided then to apply W. Edwards Deming's teachings like the PDSA, the 14 points for management, the chain reaction, and the system of profound knowledge (see Appendix 1) to help them change their current behaviors. Their recipe for success in the long term is "We basically let the innovator play with no fear."

Similarly, in another interview,[49] Jeff Schiefelbein, chief culture officer of Company 5, which named as one of the most "Innovative Companies" by Entrepreneur.com, said: "The key to success is to give a free pass for people to be creative, as long as they have in mind the purpose and vision of the company. Sometimes formalities, structure, and fear tamper the spirit. For instance, some of their employees got more innovative than ever when they started to work from home, given that they were less distracted."

. . . .

If autonomy is promoted,
remote work can boost innovation.

–Jeff Schiefelbein

. . . .

Train Them and Give Them Tools

Contrary to what you may think, and based on what I have exposed so far, being innovative is not a genetic trait. It is more a habit that you can develop by practicing it every day using different methods that help you connect ideas. These include design thinking, mind mapping, brainstorming, deconstruct-reconstruct thinking, Triz, and other problem-solving techniques. You can turn it into part of your day-to-day, too. Once you acquire the right habits, you can innovate everywhere, even at home. Train your team to become innovators.

Design Thinking, a Method to Drive Innovation

While innovation creates something disruptive, it doesn't mean it is something totally crazy. You may innovate by using something old in a different way. Innovation can be disciplined by using techniques that help people to think, imagine, observe, and connect things.

One of the methods I like the most to drive innovation is design thinking. It is a method of creative problem-solving that focuses on customer empathy. While it is usually applied to develop new products and services or make current products more appealing, the ideation process can be used to solve any problem. In the current unpredictable climate, design thinking can help companies address change more empathetically and collaboratively.

Three keys of design thinking focus on:

1. People: Prioritizes empathy and the human aspect of the solution (a human-centered approach)
2. Processes: Ensures the idea is technically feasible
3. Business: Ensures the idea is financially viable

Through a continuous trial-and-error process, ideas are tested quickly to ensure these three aspects are covered.

There are different ways to approach design thinking. The simplest way is to divide the process into three stages: exploration, ideation, and implementation.

Exploration Phase: Observe and Listen

A multidisciplinary team made up of people from varied backgrounds, such as designers, accountants, engineers, or psychologists, observes human behavior, considers a problem they have, and mulls a solution.

The team should have no more than 10 members; the more diverse, the better, with different backgrounds, ages, cultures, and seniority. Define together the major challenges facing the company, from high costs to employee engagement or a product that is now obsolete in the post-pandemic world. Then investigate what's behind the problem. Explore the suppliers, the customers, and the processes. Look at the details. For example, look into how customers have changed their behavior. Are they buying another product? Are they buying it differently? Ask questions. Take pictures, record answers, and meet with the team to analyze the findings.

Ideation Phase: Brainstorm Multiple Ideas

In the second phase, ideation or creation, think of how the team can solve the problem. This is where the tool SCAMPER is very useful. As per *The Innovation Answer Book* by Teresa Jurgens-Kowal,[50] SCAMPER is an acronym used to trigger alternate associations of existing solutions in addressing a problem. Use it to find different approaches to the problem. It can be like virtual brainstorming.

Substitute: Can you substitute or exchange parts, materials, or components of the existing solution? Many companies, for example, are replacing permanent human resources and employee development departments with consultants to reduce fixed costs and improve training quality. Other companies are taking the opposite approach and are using current employees to do jobs that would otherwise go to contractors in order to reduce project costs.

Combine: Can you combine different steps or processes? Telehealth is a solution that combines experts, new technologies like Zoom or Facetime, and apps to make health checks more affordable and accessible. Another combination in manufacturing is total productive maintenance (TPM), which combines the skills of the main-tenance department and operators to prevent problems and reduce idle hours.

Adapt: Can you adjust a specific task or product for better output? Many ideas in a company were ideated during informal conversa-tions. Working remotely, that kind of interaction is less frequent or not possible. Many companies have created virtual spaces where employees can meet. From an informal breakfast on Zoom to Slack channels or specific tools like StarMeUp, these intimate connections can still be promoted by making small adjustments to the current communication process.

Modify, minimize, or magnify: How can you adjust the whole process? Two-hour meetings online are more disengaging than face-to-face gatherings. Can you minimize them to 30-minute meetings? Or can you turn four-hour learning modules into 15-minute smaller pieces that can be delivered online?

Put to another use: Use products for another purpose, recycle waste, or choose a different target market. Let's take Airbnb. During the pandemic, its booking system was almost obsolete because nobody could travel. Can Airbnb pivot its system to provide home-office spaces for parents who need to focus outside the home but without the expenses of an office space? Can single people rent rooms or desks during the day through the same app?

Eliminate: Can you remove parts or eliminate unneeded resources to improve a process? How many things were you keeping that now seem irrelevant? Have you found yourself noting you didn't need broken tools, old-fashioned clothes, or duplicated steps? Use the 5S method (sort, set[51]—learn more about the method in order, shine, standardize, sustain[52]—Chapter 12) to sort needed from unneeded. Now scheduling for doctor or any type of appointments has gone online as well, eliminating the process of waiting or sending emails back and forth discussing the best time for a meeting.

Reverse: Rearrange parts or reverse the process. The command and control process would require the C-suite to come up with the solutions for problems. Now, agile teams require that solutions come from the bottom up. The employees are the ones who know the processes better. Why not let them figure it out? It is the first change to successful design thinking.

Implementation: Iterate Until You Solve the Problem

And the last stage is implementation. At this point, you should have tons of ideas. Ten team members by seven SCAMPERs, that's around 70 ideas minimum! Prioritize and test these ideas quickly. Use pilot teams or focus groups to try them out and see what works best. If needed, contact a coach to help you facilitate meetings to make them more effective. Iteration is part of the process, so implement them right away and make adjustments as needed.

In Japanese, to write the word "crisis" they use two signs: danger and opportunity. Maybe this is the time for your company to fight the threat and find creative ways to unleash this opportunity.

Exploring New Technologies

Advancements in artificial intelligence (AI) and the Internet of Things (IoT) have enabled a number of technologies to drive change and innovation. Singularity University applies frameworks such as design for exponentials, future forecasting, narrative-driven innovation (using science fiction storytelling), exponential leadership, and organizations to develop solutions, says Rob Nail, a Global ambassador, associate founder, and former CEO of Singularity University. "Artificial intelligence, machine learning, hyperspectral imaging, gene sequencing, and CRISPR are just a few tools that every exponential leader should understand and possibly have in their tool chest."

Worker Safety and AI

AI has the potential to keep workplaces healthier and safer through personal protective equipment (PPE) detection, safety zoning, and thermal imaging capabilities.

For instance, companies can utilize high-definition cameras that send a gentle reminder to employees via their wearables to keep six feet away and put on or replace their PPE. They can also send notifications to the manager, enabling them to identify staff members who may require additional coaching.[53] Thermal imaging capabilities can capture employees' temperatures throughout the day, detecting possible fevers and alerting management.

Amazon[54] has developed new automated staffing schedules that use sophisticated algorithms to rotate employees among jobs that use different muscle-tendon groups to decrease repetitive motion and help protect employees from musculoskeletal (MSD) risks. "We dive deep into safety issues. For example, about 40% of work-related injuries at Amazon are related to MSDs, things like sprains or strains that can be caused by repetitive motions. MSDs are common in the type of work that we do and are more likely to occur during an employee's first six months. We need to invent solutions to reduce

MSDs for new employees, many of whom might be working in a physical role for the first time.

Human Connection in the Metaverse

The connection will continue to evolve faster and faster over time. Soon we will be collaborating in 3-D spaces in the metaverse without even leaving our house. The metaverse is a combination of multiple elements of technology, including virtual reality (VR), augmented reality (AR), and video, where users are "live" within a digital universe. Mark Zuckerberg, Meta CEO, announced in October 2021 that the company was getting ready to make this happen. In a letter[55] he said, "We are at the beginning of the next chapter for the internet." Meta is developing new technologies that will help people connect, play, explore, and collaborate in this metaverse, bringing the touch, the missing senses, to the virtual world. It will allow doing things that we couldn't even do when working at the same location: it will be "an embodied internet where you're in the experience, not just looking at it."

The metaverse will bring challenges. Being able to teleport instantly as a hologram will make it challenging to maintain privacy and opportunities, such as avoiding traffic jams and reducing your carbon footprint. Everyone will be able to build this new reality, which is why Meta is also offering free courses such as the Spark AR certification to help people and organizations prepare for what's coming. What will not change is change; its the new normal, and we all will be fine as long as we strengthen the skills that make change part of our day-to-day.

We Culture Tool:
The M3 Change Management Plan

To facilitate change, I developed the agile change management M3 method. Everyone in the organization who is impacted by change or implementation should be involved in the M3.

The M3 method has three main stages: map the context, manage people, and master change (see Figure C3.2).

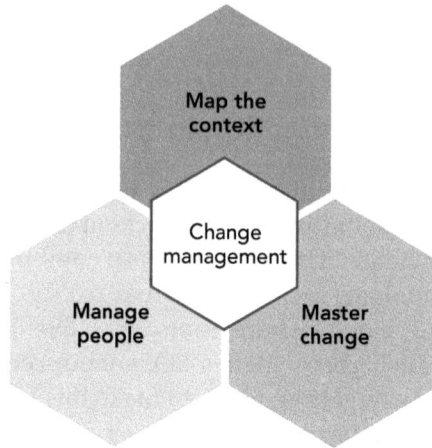

Figure C3.2   Main stages of the M3 method to achieve cultural change.

The path to success starts with the first map of the M3 model: map the context of the change by understanding where the company is today, exploring the entire employee experience, determining a clear direction, and identifying the gap involving every member and every organizational process. Then the management of change will include training personnel, coaching individuals as needed, testing the proposed actions, adjusting current processes, and managing the potential risks. Finally, to ensure the culture is sustained over time, foster and welcome continuous adjustments and reward team initiatives and engagement, but not just one time -- from now on. Define a system that normalizes and even reinforces constant change.

The M3 model has 12 steps throughout the three stages (Figure C3.3).

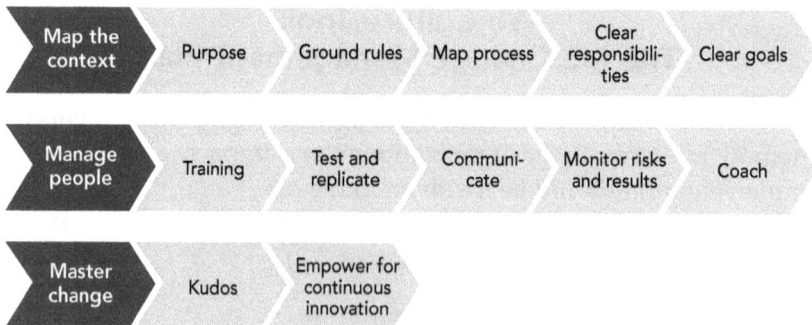

Figure C3.3   12 steps of the M3 method.

Map the Context

The purpose or "picture done": "Picture done" means ensuring everybody knows when the change has been successfully implemented, which may include showing results in a pilot test or in other companies or locations. Establish and communicate the main reasons for the change and the expectations. The purpose is key to engaging the stakeholders in the change and getting their buy-in, which is crucial for project success.

Defined ground rules: Define the 7Rs, that is, the rules of the road that will guide the 12 CARE behaviors. This will provide a scope for the teams to ensure alignment and commitment, such as be transparent, all ideas are good ideas, or the sort.

Process mapping: Many processes, workflows, and policies across the company may change. Map the current processes and relationships across activities, compare them with the desired state, identify the gap, and prioritize where to start. Define an action plan.

Responsibilities: Define clear roles and responsibilities for the change, like sponsors, pilot teams, change agents, or facilitators, teams, and leaders. Company culture is everyone's job.

Some suggestions:

- **Board of directors:** Guide the definition and development of the desired culture, ensuring it aligns with business goals and meets the needs of all stakeholders.

- **CEO and senior management team:** Set the example of the cultural behaviors and cascade key results that prioritize culture building; support and train the leaders.

- **HR department:** Build the desired behaviors into the employee experiences to reinforce the desired culture. Provide training and coaching to all employees on the keystone behaviors.

- **Quality:** Ensure the employee experience and the customer experience are aligned, communicating the culture to the customer through the employees.

- **Marketing:** Ensure branding to the customer aligns with the purpose and the culture.

- **Middle managers:** Mimic the behaviors from senior management and incorporate them into every interaction with their teams.

- **Employees:** Align their processes with the employee experience designed by HR. Provide feedback on existing culture-building efforts and ideas for new ones. Adhere to routines and norms that interpret the desired culture.

- **Clients and other stakeholders:** Provide feedback on how the culture helps meet their own objectives so the company adds value to them.

Measurement: Define the key performance measures, or KPIs, and the ones expected after the change, such as diversity targets, number of coaching sessions delivered, or team events organized. It also includes determining timelines and budgets. Start mapping the change by doing a culture survey. It will help you identify the main pain points, the areas, and the people who are more prone to change.

Risk assessment and plan for incident management: This may also include a capabilities assessment to review any additional training needs. What can go wrong? Who may disagree with the change (board members, leaders, other stakeholders), and how can they influence the result? How is the change going to impact customers, future product development, or cash flow?

Manage People

Training: Organize training sessions company-wide. Develop training sessions for leaders, middle managers, front-line users, and all the groups involved in the change, as they have different views and needs. One-time training sessions will not suffice. The CARE behaviors need to be communicated from the beginning of the company journey, during the onboarding, and when developing the employees, and reinforced in different stages of the employee experience. Augmented reality scenarios, self-paced training online, role plays, pictures, and videos can help visualize how the new behaviors feel.

Test and repeat: Start the change management with a pilot team implementing the new system, or what is called "low hanging fruit." You can get quick wins by focusing on small problems that generate a big cash impact. Start with a team that can show quick results solving issues that save time or money. If necessary, update the initial plan based on the test, and then repeat the experience with the rest of the teams. Allow time to practice the new capabilities onsite. Classroom or online training is important, but if the new skills are not practiced, the behaviors will not be learned. ERGs (Employee resource groups, learn more in Chapter 9) can also help practice new behaviors in small and private settings. Meetings, one-on-ones, events, and all types of interactions should demonstrate the behaviors. Identify "just do its" or "low hanging fruit" to start the change and show quick wins.

Coach and facilitate change: Provide specific one-on-one help to those teams that are laggards. Individual coaching is key to reducing stress; increasing acceptance, accountability, and quality; and preventing the teams from getting stuck. Coaches can be external or internal. External coaches help facilitate change by not being biased with the "as is." They can facilitate meetings, address specific concerns, and support leaders during challenging situations that they are not managing.

Change ambassadors are also important, as they are internal influencers. They may be formal or informal leaders, such as employees who have the soft skills to influence behavior and become models that other employees naturally follow. The change could be facilitated through team workshops, retreats, team-building activities, meeting facilitation, and incident management to tackle barriers or delays.

Monitor metrics, verify progress, and identify variations: Build a visual scorecard with KPIs that allow everybody to see the progress and organize daily (or as periodic as possible) stand-up meetings to communicate challenges, bottlenecks, and ideas for improvement. Internal and external audits, deep dives, or assessments may also be helpful to ensure compliance. Analyze and coach as needed when there is a variation from the expected journey.

Master Change

Share quick wins consistently and reward them. Sharing best practices, testimonials, and quick wins helps foster the changing mindset, reduce fear, keep the momentum, and increase innovation.

Empower people to innovate continuously: structure a system to continue capturing ideas for improvement on new policies, processes, and flows. The We Culture is alive. It will never be static; agile behaviors will continuously adapt to the changes in the environment to keep the organization excelling over time.

Hands-on 3.1

Think about at least three challenges you have recently overcome, such as a productivity problem within a team or a brainstorming session that did not generate new ideas. Maybe write down those challenges and remember what you learned from them. What would you tell your younger self to react better next time?

· ·

Exercise

Think about any recent challenges, and also share your learnings or the challenges you have been able to overcome.

	Challenges	Learnings
	_____	_____
	_____	_____
	_____	_____

Summary

Increase innovation by exploring, sharing, and connecting the dots.

Recap ①
Define a shared purpose and core values.

Recap ②
Nourish your network.

Recap ③
Normalize change and new technologies.

Reflection Time

Take five minutes to think about three highlights from the dimension of CONNECTION. Write them on your note pad or the action plan available on the We Culture app.

Three Highlights

_____

_____

_____

References

43. Peter Scholtes, _The Leader's Handbook_ (New York: McGraw Hill, 1998).

44. Steve Jobs, https://www.youtube.com/watch?v=UF8uR6Z6KLc.

45. Peter Diamandis, "True Breakthroughs = Crazy Ideas + Passion," _Diamandis_, 2017. https://www.diamandis.com/blog/true-breakthroughs-crazy-ideas-passion.

46. McKinsey & Company, "Why Do Most Transformations fail? A Conversation with Harry Robinson," _McKinsey & Company_, 2019, https://www.mckinsey.com/~/media/McKinsey/Business%20Functions/Transformation/Our%20Insights/Why%20do%20most%20transformations%20fail%20A%20conversation%20with%20Harry%20Robinson/Why-do-most-transformations-fail-a-conversation-with-Harry-Robinson.pdf.

47. Eric Schmidt, _Workspace Google_, https://workspace.google.com/intl/en_in/learn-more/creating_a_culture_of_innovation.html.

48. Luciana Paulise, "The Power of Innovation in Family Business," _Biztorming_, 2018, https://biztorming.com/2018/12/05/the-power-of-innovation-in-family-business/.

49. Luciana Paulise, "Best Practices for Engagement and Productivity while Being Remote," _Medium_, 2020, https://medium.com/@luciana_19373/best-practices-for-engagement-and-productivity-while-being-remote-9a38151a478e.

50. Teresa Jurgens-Kowal, _The innovation answer book_ (Houston, TX: GNPS Press, 2019).

51. Luciana Paulise, _5S Your Life: Stop Procrastination and Start Self-organization_, 2020.

52. Luciana Paulise, *5S Your Life: Stop Procrastination and Start Self-Organization*, 2020.

53. Mark Bula, "As Factories Reopen, What Role Can AI Play in Worker Safety?" *Industry Week,* 2020, https://www.industryweek.com/technology-and-iiot/digital-tools/article/21137802/as-factories-reopen-what-role-can-ai-play-in-worker-safety.

54. Ben Gilbert, "Jeff Bezos is about to hand over the keys of Amazon to a new CEO. Read his final letter to shareholders right here," *Business Insider,* 2021, https://www.businessinsider.com/amazon-jeff-bezos-final-letter-to-shareholders-as-ceo-2021-4.

55. Meta, "Founder's letter, 2021," *About FB,* 2021, https://about.fb.com/news/2021/10/founders-letter/.

Part 2:
ATTENTION

The only way to reduce errors by 70 percent was to make every single employee, in effect, a quality assurance auditor. Everyone had to take responsibility for catching mistakes.

–Tracey and Ernie Richardson[56]

T he second dimension of the CARE skills to develop a We Culture is attention to details in the present moment, being intentional and being highly purposeful in everything we do. By dividing tasks among team members and focusing our attention on fewer things, we achieve better results together.

Attention is connected to your brain and rational thinking. It is about being present, listening to your team, and offering undivided focus. As per the Ph.D. expert in attention Amishi Jha,[57] "what you pay attention to is your life." Our brains are built for bias; we cannot treat all information in the same way. We purposefully need to select what receives our attention.

In the connection dimension, you opened your view to the context, to see the big picture and connect the bits of information you get intuitively. Then, in the dimension of attention, you started focusing more on the details and results, and finding cause and effect relationships that make sense. Usually, companies focus on numbers to evaluate results. Nevertheless, too much focus on numbers doesn't allow you to see the context; you can miss perceiving other opportunities. It has its downside and is driving companies to become cold and frustrating for employees.

A great example was that during the pandemic, many companies that had financial issues decided to lay off employees, reduce benefits, and focus more than ever on numbers: push to increase sales and performance and decrease costs. When the economy started to get back to normal a year later, in April 2021, the result was "The Great Resignation." Employees decided to leave those companies before getting fired, resulting in the highest departing rates ever.

What they couldn't measure was the impact those decisions had on the employees who were staying: dissatisfaction, frustration, and disengagement.

W. Edwards Deming, a famous American statistician and one of the founding fathers of the quality profession, during his conferences and books starting in 1950, he would urge companies to avoid managing by the numbers.. He would say "inhibitors to quality and productivity have crept in, among which are: emphasis on the quarterly dividends, short-term planning, creative accounting, manipulation of assets, management by the numbers, business on price tag with short-term relationship, unfriendly takeover. [...] the most powerful inhibitor to quality and productivity in the Western world is the so-called merit system or annual appraisal of people.

What it does is destroy people. Destruction of the people in the company leads to the destruction of the company."

Many companies still simply trust the numbers, forcing individuals within a team to be ranked to define salaries, offering individual incentives, and using annual performance improvement programs (aka PIP) with quotas to push employees to perform or leave. What they are actually doing is pushing employees to think for their good, a me culture, instead of promoting team collaboration, system thinking, and a long-term vision. This system penalizes people for things that could be out of their control.

The use of visible numbers to evaluate people without a thorough understanding of the psychological impact can trigger negative feelings, frustration, and fear in the long term. Are numbers bad or incorrect? Of course not, but just focusing on them can be biased. Observing people executing the processes can help you better understand the numbers.

Key Insight

The use of visible numbers to evaluate people without a thorough understanding of the psychological impact can trigger negative feelings, frustration, and fear in the long term. Observing people executing the processes can help you better understand the numbers.

The reason is that an emphasis on budgets, problem-solving metrics, and analytics keeps our brains more in the negative emotional attractor (NEA) state. We need NEA to solve problems, analyze things, make decisions, and, especially, be able to focus. While the NEA is required to move a person from vision to action, a person must spend significantly more time in the positive emotional attractor (PEA) state to achieve sustained desired change.[58]

A person's NEA activates different brain networks and triggers hormones that activate the sympathetic nervous system and, thus, the fear and anxiety associated with the human fight or flight response.

Thought-proving questions like "What is important in your life" can awaken a person's PEA, activating parts of the brain that trigger hormones, the parasympathetic nervous system, that are associated with emotions of joy, gratitude, and curiosity. The PEA is the neurological, hormonal, and emotional state in which we are more open to ideas, other people, and learning.[59]

Suppose instead we could pay more attention to what is behind the numbers and help employees understand their own results. We could help them understand what quality means for the products they are designing, so they can help build quality into the product too. In that case, we can better help employees model positive behaviors with a positive and growth mindset, avoiding anxiety and fear.

For instance, Toyota makes its employees initial the parts they make. This process makes them pay more attention to build perfect components, feel proud, and be accountable.

Employees may engage rationally or emotionally, but emotionally is sustainable and more productive for both the company and the employee in the long term.

People must get back to reprioritizing what needs focus and spend more time analyzing those facts and their context. To avoid missing details, you need a team to focus attention on different things, and then you get together to compare and find solutions. Sharing the burden helps reduce stress and anxiety and achieve better results.

There are three desired behaviors in the dimension of attention that will drive better quality:

- Pushing decision-making to the front line

- Understanding the why behind goals and performance metrics

- Managing risk by sharing, observing, and listening

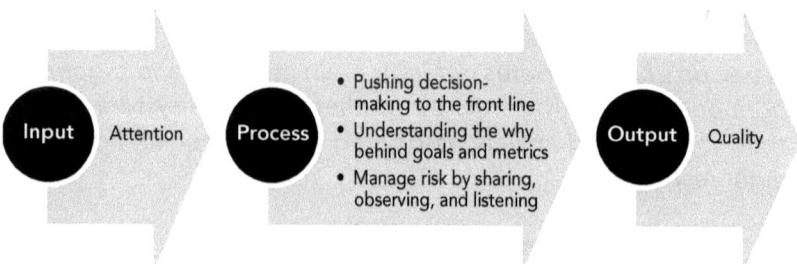

Figure P2.1   The process of decision-making in the attention dimension.

References

56. Tracey and Ernie Richardson, *The Toyota Engagement Equation*, Indian edition (India: McGraw Hill Education, 2018).

57. Amishi P. Jha, *Peak Mind: Find Your Focus, Own Your Attention, Invest 12 Minutes a Day* (HarperCollins Publishers, 2021).

58. Richard E. Boyatzis, Kylie Rochford, and Scott N. Taylor, "The Role of the Positive Emotional Attractor in Vision and Shared Vision: Toward Effective Leadership, Relationships, and Engagement," *Frontiers in Psychology*, 2015, https://www.frontiersin.org/articles/10.3389/fpsyg.2015.00670/full

59. Richard Boyatziz, Melvin Smith, and Ellen Van Oosten, *Helping People Change* (Brighton, MA: Harvard Business Review Press, 2021).

4

Skill #4: Push Decision-Making to the Front Line

The Toyota Production System or "lean manufac-turing"—relied on pushing decision-making to the lowest possible level. The NUMMI plant was one of the first Toyota plants to implement this in the United States.

Decision-making brings about a sense of control, which not only inspires but also drives innovations. As a NUMMI employee says in her book "people need to know their suggestions won't be ignored, that their mistakes won't be held against them. And they need to know that everyone else has their back. The decentralization of decision-making can make anyone into an expert.

In Toyota they are trained to lead and learn. They practice what they learn by coaching their employees. The key is "lead not from a position of power, but an empowering position."[60]

D ecision-making brings about a sense of control. If the decisions are all made by the same person, then the engagement of the other team members suffers.

As Deming would say:[61] "The obligation of any component is to contribute its best to the system." Each team member needs a clear role and responsibilities for the specific process or project.

In a We Culture, each individual has the responsibility to do their role and the freedom to do it how he or she decides to do it to achieve the best possible result for the organization. Deming was hired in Japan after WWII to offer statistical training to quality managers and leaders, engineers, and operational workers to enhance their ability to make decisions, measure performance, and drive change at the right time.

His solution was to instill pride in workers by training them and giving them the opportunity to make more decisions. Companies like Toyota received that training, which helped change the organization. Its own employees developed tools for quality management that are used even today, such as 5S, SMED, the 5 Whys, and more. These tools converted operational workers to knowledge workers capable of evaluating and correcting their own work.

Make the Decisions Where the Action Takes Place

Do you have a problem? Ask the person closest to a problem how to solve it. The nurse, the operator, the sales person — whoever deals with the problem every day. Teach these people to measure onsite so they are better prepared to make or support their decisions. A nurse may track how patients react to a medicine better than a doctor if he or she is the one taking care of the patient all day.

Decentralizing decision-making helps people make better and faster decisions while inspiring the workforce to feel important and needed. And at the same time, leaders can have more time to focus on strategic decisions instead of tracking the day-to-day operations.

Key Insight

Decentralizing decision-making helps make better and faster decisions while inspiring the workforce to feel important and needed.

With the right guidance, empowerment, and training, giving employees the authority to make decisions in their work drives multiple benefits. The organization can take advantage of everyone's ideas by pushing decision-making to whoever is closest to the problem in each situation.

*In a production plant operation, data
are highly regarded, but I consider facts
to be even more important.*

—Taiichi Ohno[62]

Data are recorded measurements resulting from an instrument, detector, gauge, or human observation. There is always a degree of error in measured data. Facts, on the other hand, are the result of direct observations "in situ" by an expert, where you can gather information about the context.

Me Culture Behavior: Paying More Attention to the Problem, Task, or Numbers, Not the Person

Richard Boyatzis, in his book *Helping People Change,*[63] says, "Where is the attention? When you pay more attention to the problem or task, and not the person, it is an obstacle to lead and help people change."

A focus on numbers makes leaders start with people objectification, worrying about what they do, not how they do it. The focus is not on helping people but on finishing tasks. Leaders are considered heroes. They are "supposed" to know everything and are thought to be the only ones capable of making the decisions. Instead, they should help to remove obstacles and allocate resources.

As a leader, leave the almighty hero role and take on the role of an observer. Then, help your team members make up their minds by pushing the decision-making to them.

Key Insight

As a leader, take on the role of observer and help your team members make up their minds by pushing the decision-making to them.

If you give people an opportunity to feel a sense of control and let them practice making choices, they can learn to exert willpower.

Create a safe space to make mistakes simultaneously so they are not afraid to take responsibility. You can do this by fostering transparency and helping them analyze risks. Remove the obstacles so team members can find the answers. It is *not* your job to have all the answers, nor to make all the decisions.

As Deming used to say, "96% of the problems are originated by the system (defined by the management), and the other 4% is due to a special cause." Unfortunately, leaders only get to know about the 4% and not the other 96%, unless they pay attention, listen, and ask questions.

> *96% of the problems are originated by the system (defined by the leaders) and the other 4% is due to a special cause.*
>
> –W. Edwards Deming

This happens because most leaders are too busy analyzing and making decisions to see what's really going on.

Henry Mintzberg, the great business thinker, told me in an interview that "Leaders need to change. They are not attuned to what is going on in the business. They don't live and breathe and care about the details; they are managers at a broader level." He said, "I think that management and leadership, the two have to go together. Nobody wants to be managed by someone who has no leadership qualities, but nobody should be led by someone who has no management qualities, because they don't know what is going on. Management is about what is going on."

"You don't make people do it, you provide them with the conditions, the infrastructure, and the knowledge to do it. If they know they are being listened to, then they want to come up with ideas, they want to help. Nobody wants to go to work and not care. You are getting paid, but it is nicer if you care, and it is nicer if people care about you. And it is simply about using all the talent that people have."[64]

Ikea, for example, sells unassembled furniture. The idea came from a worker who tried to put a table in his car and it didn't fit, so he had to take off the legs. So somebody observed, listened to the employee's need, and said, if we have to take the legs off, our customers probably do too. That is the power of paying attention.

Still, knowing what's going on doesn't mean you have to micromanage to know and control everything. Knowing what's going on is about being on the ground and listening to the employees, and it's about laying the foundations to catch problems before they become big problems. Being on the ground is how you learn about the other 96% you are not being told about.

The Socratic Walk

In order to "know what's going on," leaders should learn to see with the employees' eyes, and they can only do that by talking to them or by visiting the actual place where the work is being done. The Japanese call this place the "gemba." Lean methods always urge the leaders to "go see, ask why, show respect" like Fujio Cho, Toyota chairman, would say.[65] The idea is not to go to find mistakes or to explain how things should be done. Leaders can learn more by asking than by dictating.

> *Go see, ask why, show respect.*
>
> —Fujio Cho, Toyota chairman

I call it the "Socratic walk" because it is the same practice Socrates proposed in his teachings. As per Wikipedia, the Socratic method[66] is a form of a cooperative argumentative dialogue between individuals, based on asking and answering questions to stimulate critical thinking and to draw out ideas and underlying presuppositions. Socrates would say, "I know that I know nothing." That is the kind of approach We Cultures should have. Be open to learning: practice the student mindset. As an observer, you can train your team members to make up their minds instead of simply waiting for an instruction. Communicating updates is critical, but active listening is even more important. Encourage people to share information by democratizing digital folders, documents, and scoreboards. Provide access to trends, customer complaints, and future trends so your team is better prepared to make decisions. Use pictures, images, and videos to facilitate understanding. Create a safe space to make mistakes.

The 1-800 Contacts call center managed to push decision-making down to the organization. After four weeks of onboarding training, team members are empowered to do what they need to do to take care of a call without asking their team lead or following a script. They are even given a budget to spend in WOW moments (memorable experiences for customers when purchasing a product or service) for their clients when they need to fix a problem or help them in any way possible. Strengthening employees' autonomy and empowerment to decide how to work makes it easier for them to work remotely and easier for leaders to manage their teams.

1-800 Contacts CEO, John Graham, affirmed in an interview, "I am a great believer that ideas come from the bottom up," as several ideas from the employees have turned into commercialized products.

. . . .

Watch John Graham interview at
https://youtu.be/ekEqtFVbiwo or download
the We Culture app to watch more interviews.

. . . .

Help Employees Make Decisions

Transitioning to make the employees make decisions is a challenging journey. Most employees are used to leaving every decision to leaders or postponing decision-making (aka procrastinating). They may have been punished in the past for making a decision, so they procrastinate to avoid receiving negative feedback again. Both team members and leaders need to be trained and their habits must be rewired to change these behaviors and, in the end, change the company's culture.

Key Insight

Both team members and leaders need to be trained and their habits rewired to change these behaviors and, in the end, change the company's culture.

When some leaders are told to push decision-making to the front line, they simply stop making decisions and hope the employees will make the right choices. Unfortunately, that will not help either, because again, employees are not accustomed to making the decisions; they lack self-confidence and trust.

> *If I'm making decisions someone else can make, it's a mistake.*
>
> –Gen. Stanley McChrystal[67]

Speaking again of rules and principles, the main reason they exist is to help employees make decisions. For example, the following are Google's principles.[68] If you have a crazy idea at Google, would you share it, or would you be scared to disclose it? The principle of "look for ideas everywhere" expresses clearly that it is a culture of "yes," so you should be better at trying or showcasing your ideas than keeping them to yourself.

1. *Think 10X:* True innovation happens when you try to improve something by 10 times. If you dream big, almost impossible, you may not accomplish it 100%, but for sure you will get further than you thought you would.

2. *Focus on the user, not the competition.* Everything Google does is focused on satisfying the user, not the customer.

3. *Fail fast and learn.* Launch, then keep listening to the user. When Google launches a new product, it does a "soft launch" first within a small audience and asks for feedback. The key here is to help employees overcome resistance to failure by fostering trying and experimenting and not penalizing mistakes.

4. *Share everything you can.* The mission and objectives are shared everywhere for everyone.

5. *Use data, not opinions.* Opinions or gut feelings are not enough. If you have data, you can challenge anyone, even a manager.

6. *Look for ideas everywhere, as long as they are aligned with the purpose.* It is a culture of "yes" more than "no." Ideas can come from anyone; it doesn't matter their position or hierarchy, as long as the idea is supported by data.

7. *Hire the right people.* Google has a very robust hiring process. More than knowledge or expertise, it focuses on hiring "people who are great at lots of things, love big challenges, and welcome change."

8. *Use the 70/20/10 model.* This says that 70% of the employee's time is dedicated to the core business, while the employee can use the rest of the time for projects he or she is passionate about; 20% of the time can be used for projects related to the core business, and 10% to projects unrelated to the core business. They think that "In the long run, a few of those unrelated 10% ideas will turn into core businesses that become part of the 70%."

Three of them have to do with measuring and pushing decision-making to the front line: Fail fast and learn, share everything you can, and use data, not opinions.

We Culture Behaviors

To move decision-making to the front line, the following six steps will be helpful.

Experiment, Learn, and Repeat

A characteristic of this VUCA (volatile, uncertain, complex, and ambiguous) context is that there is not much time to make decisions, even when we have access to more information than ever to find the best answer.

Charlene Li suggests, "Instead of collecting as much information as possible before making a decision, define what I call the minimally viable data needed to decide to move you forward just one step. As soon as you know that one option is a little better than the other, make the decision and try it out. Worst-case scenario: it doesn't work out, and you can go back and try the other option."

Just as doctors employ triage to figure out which patients they should address first to maximize the number they can help, you need to do some triage on what you'll pay attention to right now. Like the Pareto rule states, focus on your most critical 20% first.

The way employees can get more information is by experimenting. So, allow them to go and test, learn, and repeat the loop.

Help your employees overcome resistance to fail by fostering the acts of trying and experimenting and not penalizing mistakes. Teach them to launch, then keep listening to the user. Soft launches are good to launch a product within a small audience and ask for feedback.

Use the minimum portions of information to test small things frequently, daily, or weekly, instead of big projects once a year. When you see how apps are designed, they are updated almost every day.

Don't be afraid for you or your team to fail; as soon as you can test it and improve it, you will be fine.

Again, remember this loop: experiment, learn, and repeat. Don't get frustrated. Be patient.

It is worse to get paralyzed than to fail.

Share Everything You Can

Every individual grows their expertise in a specific process or matter; why would a leader think he or she knows more than that individual? Leaders need to be open to the loss of control and delegate the authority to those who possess the expertise. As mentioned before, put yourself in the position of observer and learner, and see what the team can find out.

Encourage people to share information by democratizing digital folders, documents, and scoreboards. Provide access to current trends, customer complaints, and future trends to better prepare your team to make decisions. Use pictures, images, and videos to facilitate understanding; you will learn more about visual management from Skill #11, how to build team routines.

At the core of sharing information is the idea that problems should not be hidden.

- Share the mission and objectives everywhere for everyone. Ideas can come from anyone; it doesn't matter their position or hierarchy, as long as data support the idea.

- Don't avoid challenges to make employees feel safe. Instead, challenges will be well received as long as they are not threatening. In fact, Google's environment is highly demanding and innovative. It is precisely in this type of environment where the ability of people to feel comfortable taking risks, deciding fast, and participating is most needed. Biology shows that an unthreatening, challenging environment helps increase oxytocin in the body. This hormone is a neurotransmitter that is involved in behaviors related to trust, altruism, generosity, bonding, caring behaviors, empathy, and compassion, and in the regulation of fear, eliminating paralysis responses.

- Create an instance, such as an all-hands meetings, where the executives share results directly, and all employees can participate and ask questions (any questions) periodically, either remotely or face-to-face. The habit of transparency is developed here.

- Have the humility to recognize what you don't know.

- Recognize that customers and front-line employees can provide better answers for how to implement that vision on a day-to-day basis.

Toyota plants use the Andon cord to alert when there is a problem on the production line. A cable over the workstation activates a warning light or audio alarm, usually soft music. It alerts the entire team that someone is having an issue, and it also means that the line has stopped. Any problem or deviation is exposed, and the process is stopped until it is resolved. The team members don't pass the problem to the next internal customer. This is a way to show employees that you trust their thinking. They know when to stop and when to ask for help.

This is also the reason why Toyota plants implement the 5S methodology, which is described in my book *5S Your Life,*[69] to get rid of all the waste, dirt, and clutter that could hide issues. You can't manage what you can't see.

One-way methods to share information can also be effective depending on the situation:

- Single-point lessons provide information on what to do, how to do it, and why.

- Improvement bulletins show changes or improvements in procedures.

- Visual reaction plans show how to respond to problems effectively.

- Status boards on docks enable the logistics provider to deliver materials to the closest point of use or storage area, thereby minimizing material handling activities and material movement.

- Identification stickers can be used in several ways to help visually control the work area and help all employees understand when something has deviated from its standard.

- Work group display boards can be used for communication among all employees within a work group. They can be used to communicate audit results, safety information, and statistics, production levels, defects levels, Pareto diagrams, minutes from stand-up meetings, awards, and employee announcements.

- Newsletters, emails, intranets, and team chats like Slack are good to communicate cross-functional processes.

Use Data

Opinions or gut feelings are good to start, but they need to be backed up by data. Employees should know that if they have data, they can challenge anyone, even a manager.

For instance, gather customer input as much as possible. To provide WOW service, it's important that every employee understands the customers' needs to improve the customer experience whenever possible. Employees need to quickly get and act on customer feedback, regardless of the size of the company.

Design thinking or defining the customer journey can help you get information from the customer by observation (see Chapter 3).

Learn to Persuade

Rely on the power of persuasion. Instead of telling employees what to do, spend your time encouraging them to go and see other teams, come up with new solutions, and observe their own processes. Get them excited instead of pushing them to do what you think is best.

Let your employees make decisions and make mistakes. Don't micromanage; if they have questions, they will come to you. Just like kids, if you do the homework for them, it will be perfect, but they won't learn anything, and they will feel weak. To trust your team, find their strengths and support them. Trust they know how to do good work and observe to find out how you can help. Especially working with a remote team, be attentive in different ways considering that you cannot see how long they work or when they take breaks. You don't need to know when they take breaks, but you do need to know how you can help them better. Ask them.

Create a Safe Space

The last key to delegating decision-making is creating a safe space for employees to feel comfortable making decisions. This means it will be OK to identify errors or problems; it is not OK to hide them or avoid reporting them. That is why you want to ask questions, listen, expose reports and results, and build trust. If you provide all these means to share information—but then when you receive the information, you blame, shout, or penalize—you won't be fostering disclosure. They will prefer to stay put than to open up.

While in the past this was taboo, today it is accepted as part of the journey. Changes are so sudden and vertiginous that if a product takes two years to go to the market, it is already obsolete. Proof of this is the unimaginable fast development of COVID-19 pandemic vaccines. For leaders, as role models, overcoming the fear of failure is one of the critical skills they need to develop to facilitate a We Culture.

Don't overreact or penalize mistakes. Accept them as moments of truth; if you penalize a mistake, your employees will never feel confident again making decisions. Let them find their own way, feel the impact of the problem, and decide how to solve it better next time. Build a safe space by sharing performance metrics and results more openly. When employees are kept in the loop and understand their role in the purpose and goals, they are more engaged and have greater trust in their leaders. See mistakes as opportunities for improvement. You can see mistakes and ideas as problems you fear, or you can see them as opportunities, accept them, and learn from them. They will be there anyway, so it is all about how you deal with them.

Be Available When They Ask For Help

Don't become an abandoning leader (see Skill #10, promote autonomy) in the name of building trust. Problems don't belong to the employees solely, so let them know you are there to help and plan periodic check-ins. Trust that they will contact you when they need help. Touch base on agreed-upon schedules for as long as they need to build a safety net and help them when they need it the most. Agree on the frequency to be reachable but not become a burden.

Hands-on 4.1

Think about your work and what types of decisions you are able to make. Do you have a supervisor or co-worker who decides for you? How many of those decisions could now be solved by you? What do you need to make these changes? Do this exercise with your teammates or leaders to discover together how to make your processes more efficient and your job more satisfactory, and free your leader from some tasks.

Exercise

Think about your work, and what types of decisions you are able to make. Do you have a supervisor or co-worker who makes decision for you? How many of those decisions could now be solved by you? What do you need in order to make these changes?

You do not decide	You can decide
_____	_____
_____	_____
_____	_____

What do you need to change? _____

References

60. Tracey and Ernie Richardson, *The Toyota Engagement Equation*, Indian edition (India: McGraw Hill Education, 2018).

61. W. Edwards Deming, *The New Economics: For Industry, Government, Education*, second edition (Cambridge, MA: The MIT Press, 1994), 66.

62. Taiichi Ohno, *Toyota Production System: Beyond Large-Scale Production* (Boca Raton, FL: Taylor and Francis Group, 1988).

63. Richard Boyatziz, Melvin Smith, and Ellen Van Oosten, *Helping People Change* (Brighton, MA: Harvard Business Review Press, 2021).

64. https://biztorming.com/2017/11/27/management-leadership-collaboration-henry-mintzberg/.

65. John Shook, "How to Go the Gemba," *Industry Week*, 2011, https://www.industryweek.com/operations/continuous-improvement/article/21960725/how-to-go-to-the-gemba.

66. https://en.wikipedia.org/wiki/Socratic_method.

67. MIT Management Sloan School, "Retired U.S. General Stanley McChrystal Talks Leadership Strategy," *MIT Sloan,* 2015, https://mitsloan.mit.edu/ideas-made-to-matter/retired-us-general-stanley-mcchrystal-talks-leadership-strategy.

68. *Workspace Google,* https://workspace.google.com/intl/en_in/learn-more/creating_a_culture_of_innovation.html.

69. Luciana Paulise, *5S Your Life: Stop Procrastination and Start Self-Organization,* 2020.

5

Skill #5: Understand and Share the Why Behind Goals and Performance Metrics

In place of competition for high rating, high grades, to be number one, there will be cooperation on problems of common interest... the result will in time be greater innovation, applied science, technology, expansion of market, greater service, greater material reward for everyone. There will be joy in work, joy in learning... Everyone will win; no losers.

–W. Edwards Deming

S etting goals and measuring performance are probably some of the most challenging activities of a leader. Why is it that leaders keep trying to define arbitrary goals by themselves?

For some roles, productivity seems to be easy to measure: sales, claims processed, customers, and so on. But for other roles, it's more complicated, especially when employees work from home at least part of the time. Still, productivity is not always an effective way to evaluate performance, not even in sales teams. Productivity measures people's performance, but the actual measure is process performance most of the time. An increase in claims could mean a flawed process more than poor employee performance. An increase in sales could mean an employee was assigned the best-selling products or pushed clients too much. If the process is flawed, then the employee may be more prone to make mistakes. How can you measure employee performance without mentioning engagement and collaboration?

In the past, many leaders measured performance based on the number of hours people were in the office. Some were even very specific to control the time the employees would come into the office, the time they would leave, and their lunch time. In manufacturing, this may make more sense because the machines and equipment need specific maintenance. But is it really required for office workers? What happens if they work away from the team leader?

Some companies try to measure performance for remote workers by using apps to track mouse clicks, typing, or screen time. Again, in most cases, productivity cannot be accurately measured with these tools, let alone the quality of the work. As per a Gallup research,[70] "Many CHROs take a strong stance on the intrusive nature of these practices. In fact, some contend that these types of practices signal to employees a lack of trust and are a drain on engagement and culture."

Is time on screen a reference for them working on the right things? Are longer breaks a sign of disengagement or lack of interest? Some companies were able to measure engagement based on some behaviors tracked by their software. How can goals be connected to these measures? And what do leaders do with the information?

Current work setups, especially in hybrid workplaces, make it unreal to continue measuring employee performance the same way you would have some years ago. So can't you simply trust that employees will work the number of hours they are being paid for?

The Goodhart's law says: "When a measure becomes a target, it ceases to be a good measure."[71]

If an employee is forced to work from 8 a.m. to 5 p.m., he or she may do work that can be done in four hours in nine hours just to meet the target.

For some people, the measure simply makes them perform worst. They get so attached to what is expected that they don't produce anything out of the ordinary. The targets limit their potential.

Similar is the "cobra effect" when an incentive unintentionally rewards people for making the issue worse.[72]

The term cobra effect was coined by economist Horst Siebert based on an anecdote. The British government, concerned about the number of venomous cobras in Delhi, India, offered a bounty for every dead cobra. Initially, this was a successful strategy; large numbers of snakes were killed for the reward.

Eventually, however, enterprising people began to breed cobras for the income. When the government became aware of this, the reward program was scrapped. When cobra breeders set their now-worthless snakes free, the wild cobra population further increased.

If employees are rewarded by the number of cars they sell each month, they will try to sell more cars, even at a loss. If their rewards are organized by tiers, they will stop selling when they are far from the next reward.

Another employee is working from home. He finishes the work planned at 3 p.m., so he decides to work on a new project he has in mind and finishes working at 7 p.m. because he enjoys it and doesn't feel under pressure.

When employees get into the flow because they're not worried about the time, results can be surprising. That is how the Google Ads algorithm was born. An employee saw the problem posted on the kitchen board (remember the principle "share everything you can") and decided to stay all weekend with his team to solve it. They finished on Monday morning and then went home. This was not exactly a typical business day. Still, it paid off way more.

Me Culture: Emphasis on Numbers

Each person responds to stress and fear differently. Some are motivated by fear, others are paralyzed. The reasoning behind this behavior is that the response varies according to the context. People tend to avoid doing things that can negatively influence how others perceive their competence. Though this is a form of self-protection, it can negatively impact how people interact in a team. For instance, someone who doesn't feel psychologically safe avoids proposing ideas or speaking up when noticing a problem.

Emphasis in the United States is still on quantity, not quality.

–W. Edwards Deming[73]

Performance measured on the basis of just numbers makes people feel stressed out. As mentioned before, numbers drive the fight-or-flight response. People are afraid to fail or be perceived negatively because they don't want to risk what they have. Being the first can be good, but the consequences can sometimes be scary, as a new idea could be mind-blowing and start a chain reaction. That's why most people tend to stay quiet and avoid risky proposals to maintain the status quo and avoid being rejected or punished, especially if the company punishes taking risks.

When employees perceive that their job stability is on the line with every small mistake they make, they feel more nervous, defensive, and disengaged. Moreover, they are probably more focused on looking for other opportunities outside the company or fiercely fighting for any chance in the current company than performing as usual. In many cases, this fear could even become a self-fulfilling prophecy, where the employee is so fearful and distracted that he or she ends up not performing, hence being scrutinized even more.

What happens is that there is a lot of information that cannot be measured.

Take the following examples, for instance:

- Return on investment from sending an employee to training
- Losses to the organization due to fear created by performance evaluation
- Lost business due to poor service from a call center that has high turnover
- Increased business due to releasing software with bugs removed using better software testing strategies
- Profits due to employees making customers smile in your restaurant

What Deming, as a statistician, proposed was to focus on quality, not solely on quantity, to evaluate performance. Quality would bring pride to the workers. While he suggested this from 1950 to 1990, this is still valid today. Many companies are still focusing on numbers, evaluating people based on how much they sell, for instance, instead of how they sell.

Me Culture: Rankings

Me-centered cultures may promote competition among individual team members by measuring performance using rankings. All employees are ranked on a best-to-worst scale. Through the annual performance evaluations, they communicate where employees are in a company ranking. Whoever is at the top 5% of the ranking gets all the attention: the trips, the training sessions, the stocks, the development opportunities. The last 5% are on the verge of being dismissed. The other 90% simply do their jobs and get whatever is still on the table after the 5% get the most benefits. There can´t be a team of top performers, because every team member is ranked within a team, so every team has a top and a bottom, no matter how the members perform. How do you compare employees across different functions such as engineers, salesmen, managers, accountants, operators, or subject matter experts? How do you compare team members when they work hand in hand every day without expecting they will be fighting to be #1 instead of collaborating?

When a salesperson sells, is he or she the only one who should receive a bonus? What about the person who produced the product, the one who organized the store, the one who took the credit card, or the one who offered post-service?

Moreover, how is it possible that a system in which 90% of the employees don't receive any praise seems to be fair?

W. Edwards Deming wrote in his manuscripts at the Library of Congress:

> Many companies in America have systems by which everyone in management or in research receives from his superior a rating every year. On the basis of this rating, employees are ranked for raises, for example, outstanding, good, down to unsatisfactory. Management by fear would be a better name. This practice, by destroying people, has successfully devastated the western industry.

> The basic fault of the annual appraisal is that it penalizes people for normal variation of a system. The merit rating nourishes short-term performance, annihilates long-term

planning, builds fear, demolishes teamwork, nourishes rivalry and politics.

It leaves people bitter, crushed, bruised, battered, desolate, despondent, dejected, feeling inferior, some even depressed, unfit for work for weeks after receiving a rating, unable to comprehend why they are inferior. It is unfair, as it ascribes to the people in group differences that may be caused by the system they work in. The effect is exactly the opposite of what the words promise. Everyone propels himself forward, or tries to, for his own good, on his own life preserver. The organization is the loser.

There are many ways rankings negatively impact employee engagement:

- Employees start making decisions based on what is better for them, sometimes to the detriment of the company benefit (such as hiding mistakes or stealing ideas from others). This is because they know if someone is better than them, they get access to fewer opportunities.

- Employees don't go the extra mile. They are trying to play it safe so there is nothing to be criticized during the annual evaluation.

- Top employees get overwhelmed with the opportunities, while the other 90% crave them.

- Employees blame each other for problems instead of focusing on working together on the solution; guilt ignites the fight-or-flight response.

- Employees get the implicit message that some are more important than others. Some contribute more to the company than others, while they may be doing the exact same job. Therefore, they get demotivated, not understanding how their work is connected to the company earnings, disengaging.

- If a team has 10 employees, and the 10 employees did the same course or were part of the same projects, only two will get promoted or benefit. The system doesn't allow everyone to be successful.

This situation becomes a self-fulfilled prophecy, where the employees who are not top performers perform worse. It is not that employees are not good enough in most cases, it is that they are not engaged enough due to the lack of transparency and fairness in the performance review process. This makes sense, considering that statistics show how most employees are not engaged at work.

Hierarchies should only be used to organize work, not define who is better or worse or who is allowed to talk to whom.

We Culture: Setting Performance Expectations With Questions

After all this bad marketing, I want to clarify and emphasize that goals are not bad, as long as they are not used to punish and the why behind them is explained. They should be used to guide and clarify the purpose. The real problem should be measuring employee performance and setting goals transparently so all team members feel comfortable, safe, and treated fairly. Here we can learn from the IT industry, which has learned to measure performance in its teams: each individual sets and measures their own performance and makes it available to everyone.

Ashutosh founded an app called Sunsama to help employees manage their time and keep a daily account of the tasks they complete. He even included a feature to estimate how long you plan to work on a task and a feature to track the actual time spent. He calls this the "daily planning ritual." When you register for the app, he sends 10 welcome time-management-themed emails. For example, on email #9, he recommends, "During my daily planning ritual, I defend myself from this by planning my day and measuring my performance not by how long I want to work, but what output I want to achieve at the end of the day. If I finish up my goals for the day by 3 p.m., I go home. Since my team also uses Sunsama, they can see my output for the day and know I'm heading home because I got a lot done."

Apps like Sunsama help people be accountable for their own performance measurement (turning on the timer for each task). But the key here is that the tracking is the one helping the employee work better, not the leader. You can use this measurement to punish employees for not working the same number of hours every day, or you can use it to better understand and organize your time. The first option is managing by numbers, focusing on how much is done, and the second option is focusing on quality, how time is used.

You can set goals at the individual or group level, but the employee should set them, not the leader. Many leaders complain this is not possible because employees "don't know how to estimate" or they "will estimate more just in case." Employees need to learn to plan their tasks and be held accountable for what they estimate. Research shows that performance is higher when people are committed to their goals.[74]

Rather than giving orders, leaders in agile organizations learn to guide with questions such as: When do you think you can get this done? What do you recommend? How could we test that? Instead of stating what to do, asking questions shows employees that you trust them and helps them link small tasks to larger and feasible aspirations. They understand where the decision is coming from and the importance of their tasks, and they feel part of a larger process. This way, they collaborate more and bring more perspectives, especially from what they learn from clients.

Examples of High-Impact Questions You Can Ask Team Members

- Ask for more data and facts.

- Ask open-ended questions.

- Ask for ideas, alternatives, possibilities, or feelings.

Types of questions

Results

- What does success look like for you?

- How would you describe your current performance?

Opportunities

- What do you recommend?

- What opportunities do you see?

Execution

- What do you see as the timeline for completion?

- How can we test it?

Problems?

- What resources do you need?
- What risks do you see?

The 5 Whys

Another good use of questions is solving a problem, as we already did in the Introduction. The 5 Whys is a method used to identify the root cause of a problem; Toyota's Taiichi Ohno urged workers to ask "Why?" five times.

The 5 Whys is a root cause analysis technique that is often used to analyze a real error that has occurred.

> *The essence of being human involves asking questions, not answering them.*
>
> –John Seely Brown,
> former director of Xerox

By asking "Why?" five times, you should be able to understand a problem deeply enough to identify the ultimate root cause, such as in the example below.

5 WHYS EXAMPLE

Problem: I am being turned down for a bank loan to fund a small business.

Q: Why am I being turned down?

A: Because the bank doesn't want to finance my business.

Q: Why doesn't the bank want to finance my business?

A: Because they don't see the true potential in my business.

Q: Why doesn't the bank see the true potential in my business?

A: Because they are not getting the proper information and facts they need.

Q: Why are they not getting the proper information and facts they need?

A: Because my investment proposal is ineffective and incomplete.

Q: Why is my investment proposal ineffective?

A: Because I have not sought advice from someone who may be able to help me better communicate the value of my business proposal.

Why am I being turned down for a bank loan? The simple answer could lead to thinking that the bank doesn't like me. This is subjective, and it makes me weaker, a victim of the situation, as there is nothing I can do about it.

Asking 5 Whys helps me understand that the reason behind the denial is that I may need to look for someone knowledgeable who can guide me to improve my proposal. This answer empowers me to continue learning, looking for solutions and doing something different the next time. This is ideally what we want from team members: to look for answers and solutions by themselves.

Enlisting knowledgeable team members ensures critical aspects of the problems are identified, understood, and tackled by the experts.

Use Employee Scoreboards to Measure Performance Onsite

We Culture is always about two-way communication. Teams should have ways to display information about their processes online that help them measure and make decisions.

Boards with metrics, trends, and results should be prepared and used by the employees. By building their own boards, team members can make decisions much faster.

For example, a manufacturing company can have a dashboard showing the number of finished products, errors, and customer claims. A call center could have the number of calls, duration of the calls, number of sales, claims, results of customer surveys, satisfaction rates, etc. A technology company designing apps can see customer complaints, observe how clients use the app, measure the number of downloads, and so on.

Most companies have a disconnect here, showing the employees only part of the information, especially when there is something wrong, such as a mistake or a complaint. The employees cannot see the whole context. They cannot detect a trend, anticipate a chain of errors, or repeat positive behaviors. They receive the information late and only what the leaders consider important.

Team members should have access to as much data as possible about their own process, see in real time what is wrong, and have the power to fix without feeling guilty.

If team members can see in real time what types of complaints the customers are filing, they can propose adjusting their script or asking for help to improve their customer service.

Visual charts help both team members and management understand what's going on. According to the book *The Toyota Engagement Equation,* Toyota uses a Yamazumi chart, which is a stacked bar or cycle chart, to display measures of the manufacturing process at the workstation. The employees use it to know the wait time, walk time, process time, repair time, delay work, etc.

You can generate these charts on a computer, or use magnets, sticky notes, or even pen and paper. What works best? Simply think about how long it takes to generate the chart. If you need to wait five days to have a computer chart, it may not be helpful once you get it. You may be better doing something simple that can be updated at any time by anyone.

Scoreboards are also great for communicating work status in virtual or hybrid teams. Many companies, especially IT teams, use Trello, Asana, Jira, or Sunsama to make sure everyone knows what's going on. These tools, unfortunately, are not used much in other industries. For example, offices and call centers could work much better together if they shared their tasks and goals in online charts.

360-Degree Reviews

Another common method to ensure the performance evaluation reflects all the employees' needs, desires, and values is the 360-degree review. This review process uses feedback, anonymous or not, from an employee's subordinates, colleagues, supervisor, and a self-evaluation. The feedback is normally reviewed with the supervisor to see how the employee's work impacts others and how it can be improved or reinforced. It is important to choose individuals from different processes, levels, and departments to have a fair review.

At least once a year, it is a good opportunity to write a statement of your own performance and see if you are satisfied with it. Would you hire yourself? Would you work with yourself?

In my perspective, the most important part of the 360-degree feedback is that employees recognize that they not only work for their supervisor, but they also work for their subordinates, teammates, suppliers, customers, and other stakeholders. When employees know all the feedback is valuable, it sends an important message that everyone is valuable too, not only top management.

360-degree reviews are usually done once a year. It is costly if every employee has to receive an evaluation from 10 other employees or contacts, and the review and analysis are also very time-consuming. That's why few companies can perform them or perform them only occasionally. With a we mindset, employees can be trained on the importance of 360-degree feedback and can perform it when needed. The survey can be a link with open-ended questions, available to all employees, that can be sent to others when an employee has low performance, or maybe when the employee has high performance compared to others to understand why or to solve a specific problem. This tool is a treasure that should not be kept in the dark.

In the current sharing economy, the companies that can grow faster are the ones that practice 360-degree feedback, not only to evaluate the employee but also the services. Customer reviews are a source of direct feedback from the customer to other customers. You can rate a restaurant or a rented apartment on Airbnb or Google Maps. On Glassdoor and Blind, you can rate companies and CEOs. When you buy a car from Toyota, you receive a link to a portal when you can suggest ideas for improvement.

Setting Goals

There are many ways to help employees set their own goals. The most common ones are the results only work environment (ROWE) and objectives and key results (OKR) methods.

The company 97th Floor, a digital marketing agency, applies the ROWE method. ROWE helps the company promote autonomy as a key value. 97th Floor give employees freedom and responsibility to achieve their goals, so it's not required for employees to work in the office, for instance. Teams are organized not by the department but by customers to be able to provide dedicated designers and writers more connected to their specific needs. Annalee Jarret, an employee with 97th Floor, says, "People feel valued when they are given the responsibility to self-manage."

A wide range of organizations, including Google, use OKRs: "They are an individual's objectives (the strategic goals to accomplish) and key results (the way in which progress toward that goal is measured). Every employee updates and posts their OKRs company-wide every quarter."[75] The objectives are the big picture, and the key results are measurable.

OKRs translate the company's priorities from the quarter throughout the organization, not cascaded through a top-down command and control approach. This secures involvement and ownership.

- All OKRs are open and visible for everyone: OKRs should be for everyone to see so every employee knows the organizational objectives and metrics for success.

- They are ambitious targets that cannot always be met. If employees attain their objectives by 100%, then their OKRs aren't ambitious enough.

- Teams set the key result targets themselves, considering top-down and bottom-up suggestions. When teams set their own objectives, they will take pride in achieving them.

- The key result is measurable.

- There is no link to bonuses, and they are not synonymous with employee evaluations. This helps to avoid gaming on target levels. Employees help grade themselves on how they performed against the previous quarter.

- All employees should easily remember their own OKRs.

- Not everything needs OKRs; they are only for special areas of focus.

A good example of an OKR could be "run a marathon in less than three hours," and it would be green only if you get the goal that trimester. If you are close, you may define when it is yellow or red. Having yellows and reds in your scorecard should be OK.

In an interview, Dick Costolo, former Googler and former CEO of Twitter, was asked, "What did you learn from Google that you applied to Twitter?" He shared:[76]

> The thing that I saw at Google that I definitely have applied at Twitter are OKRs. Those are a great way to help everyone in the company understand what's important and how you're going to measure what's important. It's essentially a great way to communicate strategy and how you're going to measure strategy. And that's how we try to use them. As you grow a company, the single hardest thing to make sure you scale is communication. It's remarkably difficult. OKRs are a great way to make sure everyone understands how you're going to measure success and strategy.

Some Tips for Setting Objectives Using OKRs:

- Pick no more than three to five objectives (again, use the Pareto rule to prioritize).

- Avoid ambiguous expressions that don't push for new achievements, such as "keep hiring" or "continue doing X." Use expressions that convey final goals, such as "finish project x," "design three new products."

- Use tangible, objective, and unambiguous terms.

- Use key results to describe outcomes, not activities.

- Set measurable milestones that include evidence of completion, and this evidence should be available and credible. In agile methods, for example, tasks to be accomplished should define when it's considered done. Is it done when someone finished coding, or is it done when it has been tested and approved? This differentiation is key to ensure goals can be measured and compared relatively.

And, finally, remember not to use OKRs simply to define who did good work and who didn't, to punish or reward. Use OKRs to understand what is going on; apply other statistical methods to identify trends and causes of variations (train your team to do so too), discover who needs help, and empower team members to offer solutions.

How Would You Reward Your Team?

The rewarding process is truly important not only to avoid frustrations but also to bring attention to what really matters.

First of all, is money enough?

Awards, external recognition, and rankings can generate frustration and disengagement among employees unless they are designed intentionally to impact positively.

W. Edwards Deming was always against sales incentives. In his book *The New Economics*[77] he said, "A furniture company in Houston put their salesmen on salary, in place of commissions for sales. Result: a steady increase in sales. Older salesmen now help beginners. Salesmen no longer try to steal business from other salesmen. They now help each other. They protect the customer."

Some questions we can ask ourselves are:

- How are we encouraging the team?

- What is the reward reinforcing?

At school, my daughter would cry when she did not receive an award. At only seven years old, she just cared about the prize, so I understood why she was frustrated. She knew she had put in a lot of effort. But the problem was that she didn't understand why she didn't win or why the others were getting the award.

Some solutions for this problem?

- Let the students decide what the award is and who should receive it.

- Don't give an award at all; trust in intrinsic motivation.

- Ensure the rules of the selection process are very well defined and transparent for everyone to understand what to do to win.

- Provide prizes to teams, not individuals.

- Discuss who would be awarded and why before giving the recognition, to manage everyone's expectations during the award ceremony.

You might wonder why you would do such a complex process for a child. Because if there is pain, there must be a way to solve the issue. You cannot judge a person for their feelings, you need to find ways to eliminate the frustration, or at least be aware of it and learn how to manage it. This process can help much better.

The way you reward your team will define what they focus on. When companies provide incentives for sales to a salesperson, then only they are motivated to work harder, and they are also unconsciously pushed to sell more, disregarding quality or customer service. That can be counterproductive. "Incentives distract people from their inner motivation, so we are better off without them," said Frederic Laloux in his book, *Reinventing Organizations*.[78] "Incentives distract people from hearing their internal motivations, so it is better not to have them."

I used to work as a telemarketer selling cell phones when I was finishing my bachelor's degree, and I remember it as my worst work experience. Even though the environment, teammates, and office building were nice, working only six hours a day, I suffered from the pressure of achieving an arbitrary quota of sales per day and meeting the quota of minutes per call. Sometimes I felt the urge to push people who didn't even understand what I was selling. For some people I had to take longer to explain everything, but that was counterproductive for my total sales per day, so I had to make it as quick as possible. I was visibly nervous about talking to people during the first months. If I didn't have to meet the quota, I probably would have sold more. I actually noticed that some of the employees who were not working for the money because they didn't need it (they were young and living with a wealthy family) were #1 in sales, while those who needed the money felt the pressure even more to avoid having a basic salary.

Salaries and bonuses shouldn't try to build a gap between those who can sell and those who can't or force you to decide if you want to sell or offer great customer service. All employees should aspire to do both. If some associates are better at sales, they should be in the sales positions, while those who are not able to make sales should be doing other tasks, maybe quality assurance, or receive more personalized coaching to try to improve. But nobody should be penalized for taking the time to serve a customer better. Goals should not be counterproductive.

When I visited Zappos, for instance, I remember the employees saying that even though they sell shoes, if a customer calls asking for a place to eat pizza, the employees will do their best to help the customer find it, even when it is "outside of their job description." Their #1 value is to deliver WOW through service, no matter how long it takes.

"We companies" eliminate competition among individuals while instead fostering self-actualization, team rewards, and peer-to-peer reviews. The system should allow all employees to succeed if they all try hard.

A focus on competing with other companies can be replaced by a sharing and abundance mindset. For instance, in 2014 Elon Musk took all his patents open source to accelerate the advent of sustainable energy, encouraging companies to work on building electric cars.

We Culture: Team Incentives

Individual incentives besides the base salary generate competition among team members instead of collaboration and frustration to those who don't receive the prize, especially if the majority are in this last group. If what is sought is to strengthen teamwork, all team members should be rewarded, with the presumption that each one contributes from their strength. No one is more important than the rest.

Of course, nothing is perfect. The problem of the group incentive is brought by free riders, those who benefit from the efforts of others. It is difficult to control, but autonomous teams do not require a judge or a manager who punishes those who do not work. The same colleagues give feedback to their peers regarding the performance at the moment. As we said, they don't wait for the annual performance evaluation. There is nothing more effective than the pressure of the team to solve unfair situations.

If awards are provided based on individual efforts instead of team efforts, individuals learn to work by themselves and avoid helping others. That's how they are incentivized to work. There is nothing wrong with that, unless you want to promote team collaboration. Team awards instead promote team spirit and value helping each other. Many companies, especially Facebook, Amazon, Apple, Netflix, and Google (the FAANG), offer stocks when an employee joins the company. Many companies have started using a shared bonus, a bonus that every employee receives, as a percentage of the company's profit. Everyone can access information on the profits and losses of the company. If they have good performance, everyone earns a little more; if they lose, nobody receives anything. Transparency helps drive the feeling of justice and equity.

So instead of incentives for particular positions, companies could offer an organization-wide bonus based on sales or other KPIs, including even hourly workers.

Just to clarify, the idea of team incentives is not that everyone receives the same payment. Here is some advice:

- While there will always be ranges for salaries, there must be an objective way to move to the next range (by seniority or based on college degrees).

- Ensure payments do not depend on performance measures unless they are the same company-wide (if the company wins, everyone gets a bonus).

- Use intrinsic motivators (pride, congratulations, showcase work to other co-workers or industry leaders, payment of studies) instead of extrinsic (individual prizes for achieving arbitrary goals, such as trips or individual bonuses).

- Use team incentives.

- Make sure every employee makes enough to cover basic needs.

- Reduce the range between the highest and lowest salaries to ensure a better distribution.

- Eliminate hourly wages; give everyone a salary.

In 2018 I interviewed Jim Schrader with the company Technology Site Planners during a seminar organized by the Deming Institute and delivered by Bill Bellows (who wrote the Foreword for this book). Jim told me his experience at Tech Site: "We used to do individual performance contracts where we would pay people based on certain criteria such as billable hours. Over time we realized that was not the best way to proceed. Sometimes that would set up a motivation to work on something not critical. A customer would have an emergency, and the best thing for the customer and the company was to stop what we were doing on the billable hours' side and work on the customer's emergency, but we had created a system motivation to do the opposite. So we stopped doing that, and now it is just based on team rewards and taking care of the customer first."

We Culture: Peer Reviews

Offer opportunities for peer reviews. Similar to 360-degree reviews, many companies are starting to apply peer reviews as a method to ensure performance evaluation is more transparent and fair; in most cases, this involves replacing employee rankings and annual reviews.

Globant is a company with more than 15,000 employees. It conducted a survey in 2019 and confirmed that 72% of respondents said they were not satisfied with the organization's current feedback process in key areas like career development (73%) and job performance (70%). With this in mind, the company developed the StarMeUp OS, allowing employees to do peer reviews. With this new system, each associate has 20 stars to recognize other team members for a specific action representing company values, such as aiming for excellence, being kind to the environment, or being more inclusive.

As another example, Zappos lets its employees engage each other. They reward one another with "Zollars." Every team has a designated person to distribute Zollars. All employees once a month can give 50 Zollars to whomever they consider deserving. For example, an employee heard about excellent customer call handling, and she wants to applaud it.

Key Insight

Public recognition is better than confidential recognition. Improvement feedback, however, is better confidential than public.

TINYpulse developed Cheers for Peers for this purpose. It's an online recognition platform that enables employees to provide peer-to-peer recognition to their new colleagues. Cheers for Peers also allows new hires to tag their cheers or rewards with a company value that the organization predefines. A reduction in tagged values may show, for instance, that employees are decreasingly identifying with company values, which can be a sign of lower engagement and commitment.

Not everything that counts can be counted, and not everything that can be counted counts.

–Albert Einstein

In many cases, even an employee's salary is chosen by that employee or by their peers. In other cases, it is based on experience with the company, and increases as an employee's roles and skills or knowledge increase. For example, Semco, a Brazilian company, makes its employees set their own salaries. At Morningstar, an American company, employees set their own salaries with the help of a committee specially formed to moderate wages.

Gamification

Some companies use gamification to measure the results of the employees and improve engagement. For example, Gamifica Group created software that is a combination of challenges, scoring systems, badges, rankings, and mechanics of cooperation.

Another example is SentinelOne. When working from home, the company organized all types of well-being programs and "gamified" many aspects of its culture, creating fun competitions and interactive events such as a virtual wellness challenge.

Divya Ghatak, chief people officer at SentinelOne, told me in an interview,[79] "Everyone recorded their steps. We used Slack channels to share what each of us was doing. In Tokyo, Oregon, Scotland, people around the world were taking pictures of how they were walking. You could look at the daily life of your colleges; it was more personal than before."

More Ideas to Increase Performance Evaluation Transparency

- Base salaries on individual or team accomplishments. Companies can establish salary levels and be transparent about what is required to move from one level to the other, such as years of experience, courses taken, formal studies, or certifications. The key is not to have everyone on the same level and ensure that going from one level to another doesn't depend on making others perform worse but rather on the employee becoming better.

- Offer a share of earnings, as long as the company offers it to every employee. Some companies provide stocks only to the top performers, while other companies like Google or Facebook give stocks to every, or at least most, employees.

- Instead of blaming someone in particular, open the conversation with that person or with the entire team so together you can discover what happened and how you can resolve it. Use questions like: I've seen this trend, what do you think? What do you think happened? How can I help you solve it?

- Use monthly reports to share, analyze results, and help, not to punish. Most companies have weekly or monthly meetings to share reports, audits, and results, punishing unexpected results. If employees feel threatened to show negative results, they may be tempted to hide them, sugar coat them, or manipulate them instead of asking for help. Meetings should be viewed as a time to analyze results, share experiences, and look for solutions as a team. When more issues are safely exposed, more solutions arise.

Hands-on 5.1

a. Remember and write down the company's purpose. Think of a goal to accomplish that could benefit that purpose given your role and strengths. Then, define at least three specific actions that you can take to attain that goal at least at 70-80%. Also, think about how you can measure those actions. Be very specific on what needs to be done and when, schedule time on your calendar, or define due dates to help you finish on time.

b. Check the progress results after a month. Are metrics enough to evaluate how hard you are working on those actions to attain that goal? What other things should you consider, besides the raw metric, to find out if you were successful.

Exercise

Work with your supervisor or teammates.

Write down the company's purpose.	Write down a goal.	Define three actions to achieve the goal and its metrics.
_____	_____	_____
_____	_____	_____
_____	_____	_____
_____	_____	_____
_____	_____	_____

Other considerations to evaluate progress.
_____
_____

References

70. Ellyn Maese and Larry Emond, "What 150 Top CHROs Are Saying About Productivity After 2020," *Gallup*, 2021, https://www.gallup.com/workplace/329702/150-top-chros-saying-productivity-2020.aspx.

71. "Goodhart's law," *Wikipedia*, https://en.wikipedia.org/wiki/Goodhart%27s_law.

72. "The Original Cobra Effect," *Wikipedia*, https://en.wikipedia.org/wiki/Perverse_incentive#The_original_cobra_effect.

73. W. Edwards Deming, Manuscripts, Library of Congress, Washington, DC.

74. Edwin A. Locke, Gary P. Latham, and Miriam Erez, "The Determinants of Goal," *Academy of Management Review* 13, no. 1, https://journals.aom.org/doi/abs/10.5465/amr.1988.4306771.

75. Eric Schmidt and Jonathan Rosenberg, *How Google Works* (New York: Grand Central Publishing, 2014).

76. https://www.youtube.com/watch?v=T963AGt-nWI#t=11m52s.

77. W. Edwards Deming, The New Economics: For Industry, Government, Education, second edition (Cambridge, MA: The MIT Press, 1994), 50.

78. Frederic Laloux, *Reinventing Organizations* (Brussels, Belgium: Nelson Parker, 2014).

79. Luciana Paulise, "6 Ways to Build a Culture in a Hypergrowth Company," *Forbes*, January 11, 2021, www.forbes.com/sites/lucianapaulise/2021/01/11/6-ways-to-build-culture-in-a-hypergrowth-company/?sh=7db9a3503143.

6

Skill #6: Manage Risk by Sharing, Observing, and Listening

What makes Pixar special is that we acknowledge we will always have problems, many of them hidden from our view; that we work hard to uncover these problems, even if doing so means making ourselves uncomfortable; and that, when we come across a problem, we marshall all of our energies to solve it. This is why I love coming to work in the morning.

−Ed Catmull[80]

T his skill is about analyzing risks as a team and building a culture of seeing risks and threats as opportunities. When it is a common thought that leaders should be the ones making high-risk decisions, teams should learn to evaluate alternatives and make decisions together. The amount of information available nowadays makes it impossible for one person to make the right call. It is important that all team members know the risks and opportunities and learn how to weigh them.

Attention is not about talking. It is, however, about withholding your thoughts, letting people talk, listening, and asking questions. That's how you can understand how you can help.

In a we culture, listening and observing become more important than demanding and controlling to reduce risks.

Different organizations have different aversions to risk embedded in their cultures. Some companies that depend on innovation to succeed prefer to take risks. A good example is Tesla, which launched self-driving cars when they were not thoroughly tested, resulting in some reported deaths. Production was delayed. Still, sales of those cars went through the roof, even when Tesla could not deliver on time. In April 2020, at the onset of the pandemic, Tesla's stock began to go up (see Figure C6.1). Around the same time, the price of oil went negative at $37.63. As per the images below, April 20, 2020, was the first day in history when oil recorded negative prices. Not by chance, oil companies are on the side of aversion to risk, managing a commodity. Still, as this price change shows, even oil companies have to analyze new opportunities and gauge risks to look for new energy sources and businesses to survive in the long term.

Interactive chart

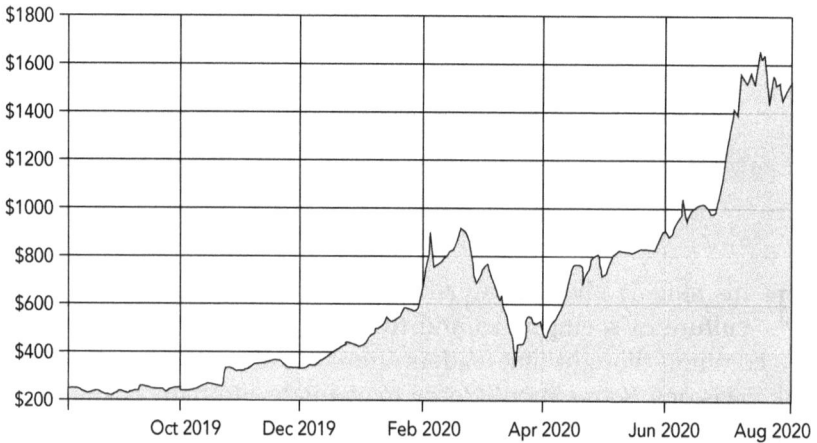

Figure C6.1   Tesla stock price by the Motley Fool.
https://www.fool.com/quote/nasdaq/tesla/tsla/.

Uncertainty requires a more careful and conscious risk analysis and a clear definition of action, not analysis paralysis or risk avoidance. It requires active listening and asking questions:

- Why are we doing this?
- What could go wrong?
- Is this safe?
- Is the customer happy?
- Could the product get damaged?

Risks are not reduced by simply avoiding them or passing the baton: the faster they are addressed, the better.

Innovative companies usually take more risks in the pursuit of innovation. They test products that may not work but also take actions to mitigate those risks. We can call this "smart risk-taking." They may try products in pilot settings, do quick turnarounds, and get quick feedback from the customer. Sometimes being slow is riskier.

Agile Thinking

The Lean Startup methodology was one of the first to promote rapid development cycles and iterative product releases to validate learning instead of waiting for the perfect product. Eric Ries says in his book, *The Lean Startup*,[81] that "Using the Lean Startup approach, companies can create order not chaos by providing tools to test a vision continuously."

Agile methods also promote testing and create opportunities for pilots, prototypes, and experimentation. The Agile Manifesto[82] has 12 principles, some of which specifically address risk management and rapid prototyping:

Principle 1: Our highest priority is to satisfy the customer through early and continuous delivery of valuable software.

Principle 2: Welcome changing requirements, even late in development. Agile processes harness change for the customer's competitive advantage.

Principle 3: Deliver working software frequently, from a couple of weeks to a couple of months, with a preference for the shorter timescale.

Principle 12: At regular intervals, the team reflects on how to become more effective, then tunes and adjusts its behavior accordingly.

While these principles mostly talk about delivering software, they can be applied to any industry. If you ignore the reference to software, the basis of these principles is to emphasize frequent delivery, continuous change, continuous reflection, and rapid response.

Failure is the New Status Quo

It must have been terrifying to be Copernicus, placing the Sun rather than the Earth at the center of the universe. Edison failed 999 times until the first light bulb worked at try 1,000. Copernicus and Edison were solo entrepreneurs, but what happens within an organization?

According to the book *The Innovator's DNA*,[83] mistakes can be productive. Still, they must have three characteristics: they must be detected quickly, must not be too big to impact the company's name, and must allow the company to learn from the error.

How leaders react to failure is also part of the reason why employees choose to avoid failure. Leaders are role models, so they react similarly if they punish failure or don't accept that they are fallible and hide mistakes. Sometimes leaders avoid proposing ideas and discourage others from doing so, fearing that it will make them look bad. As Brene Brown puts it in her book *Daring Greatly*,[84] the big challenge for leaders is "to cultivate the courage to be uncomfortable and to teach the people around us how to accept discomfort as part of growth." The fear of failure needs to be demystified to embrace innovation, change, and growth. Otherwise, you would be mining your chances to do something amazing.

> *Cultivate the courage to be uncomfortable and to teach the people around us how to accept discomfort as part of growth.*
>
> –Brene Brown[85]

Frederic Laloux, author of *Reinventing Organizations*,[86] says, "What replaces fear? A capacity to trust the abundance of life. We come to believe that even if something unexpected happens or if we make mistakes, things will turn out all right, and when they don't, life will have given us an opportunity to learn and grow."

Employees need to be supported to make decisions in their areas of expertise. This unlocks everyone's ideas and thus generates more innovation. Ideo's slogan to help employees embrace failure is, for example, "fail often to succeed sooner." This sends a clear message to employees on what is accepted and what is not. Citing Deming, we need to "drive out fear." The fear of failure makes people slower in decision-making, less creative, and less team-oriented.

Risk-taking welcomes doing things differently, such as the email I received from Amazon (see Figure C6.2) when I bought a hammock.

This means that taking risks and innovating can also help you improve your quality of service.

There is no such thing as failure if you can learn from it. In the long term, there are no failures, only steps as part of a process.

amazon

You have received a message from the Amazon Seller – Hammock Sky

---

Dear Hammock Sky Customer,

In a galaxy far, far away...

Your Hammock has been gently taken by the Relaxation Engineers from the Hammock Sky shelves with soft, velvet gloves and placed into a Mother Ship headed for Earth.

A team of 10 intergalactic minions inspected your Hammock, and polished it to make sure it was in the best possible condition before shipping.

Our packing specialist, Bob, received your Hammock in a starry desert. He lit a candle, and a hush fell over the crowd as he put your Hammock into the finest shipping box that money can buy.

We all had a parade afterwards and the whole terrestrial Hammock Sky team hiked up to a mountaintop post office and waved "Bon Voyage!" to your box, on its way to you, in the next days.

I hope you had a great experience shopping at Hammock Sky on Amazon. We sure did. Your picture is on our wall as "Customer of the Month." We're all exhausted but can't wait for you to come back to Hammock Sky!

Thank you,

Phil Bourdeau
Relaxer in Chief, Hammock Sky

Figure C6.2   Email received from Amazon seller Hammock Sky.

We Culture Tool: The FMEA

The key to a productive risk-taking approach is in utilizing methods that help people analyze what is at stake as soon as possible. The 5 Whys, mentioned previously in Skill #4, is one of these methods. Another more complex and thorough method is failure mode and effects analysis (FMEA).

FMEA helps you identify potential problems before they even occur. FMEA is a systematic approach used to identify potential issues (failure modes) and quantify the risk of occurrence. It is extensive and required in the automotive industry. The method has seven steps, first used by the United States Military at the end of the 1940s (see Figure C6.3).

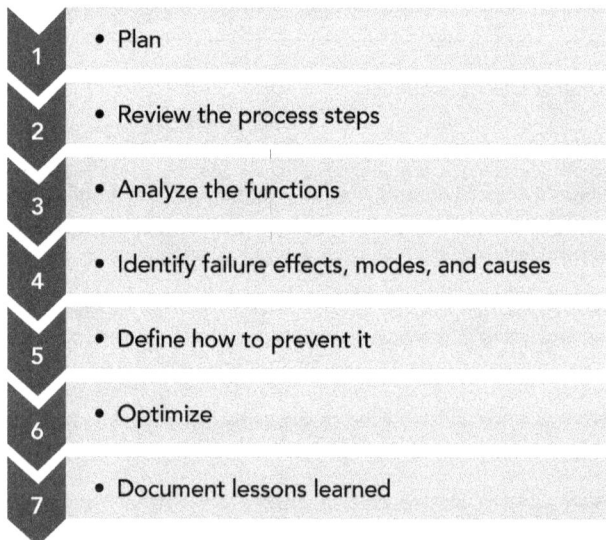

1. Plan

2. Review the process steps

3. Analyze the functions

4. Identify failure effects, modes, and causes

5. Define how to prevent it

6. Optimize

7. Document lessons learned

Figure C6.3   Steps of the FMEA.

1. Plan: Define the intent of the FMEA, the team members needed, and the time available to meet.

2. Organize a brainstorming session to describe the steps of the process under analysis.

3. Analyze and list the desired functions of the process, product, or service.

4. Identify potential failures of those functions, the effects of the failures (What is the impact on key output variables?), and the potential causes of failures (What causes the key input to go wrong?)

5. Define how to prevent the failures and recommend actions:

    a. List all the means to prevent and detect each failure effect and the different causes for each effect

    b. Use established criteria (such as customer input) to estimate the severity of each effect (severity rating)

    c. Brainstorm process changes that could contribute to the listed effects

    d. Estimate the likelihood the stated cause will occur and list all controls that currently exist that would prevent the cause from occurring (occurrence rating)

    e. Estimate the effectiveness of controls that now exist to prevent the stated cause (detection rating)

    f. Multiply severity x occurrence x detection to get the risk priority number (RPN). In the automotive industry, the RPN is not used anymore; what is used is the action priority (AP), which is high, medium, or low.

6. Optimize

    a. Prioritize the issues to work based on: high (must define suitable actions), medium (should define suitable actions), or low (can define further actions)

    b. Define follow-up actions based on the RPN or AP linked to the stated cause to be effective

    c. Assign owners to the actions listed

    d. Share progress and status reviews

7. Document lessons learned

    a. List completed actions on the FMEA form

    b. Submit completed FMEA forms to process owner for review

    c. Update when needed

Me Culture Behavior:
Reacting to the Wrong Problems

The main source of stress nowadays relates to the immense amount of information we receive. This particular skill requires us to understand where to focus our attention. Is a simple variation from the expected result a reason to worry? How can team members learn to pay attention to the right things?

Walter Shewhart determined that there are two types of mistakes that can be committed:[87]

- *Mistake 1: To react to an outcome as if it came from a special-cause variation when it really came from common causes of variation*

- *Mistake 2: To treat an outcome as if it came from common causes of variation when actually it came from a special cause*

The first mistake is called a false alarm, or *tampering* by interfering with a process too often driving a change for the worse. Simply said, it's micromanaging.

This is when a leader overreacts to rumors or checks in too often with employees. An employee may be punished due to an error he or she made that is a common cause, not the employee's fault, or due to other common causes, such as lack of resources or a flawed process. These common causes are used on multiple occasions to justify an employee's dismissal.

The same happens at home, such as when you turn on the faucet and try to find the right temperature for your shower. It's hard to get to the right temperature if you hurry to turn the hot and cold too often because you don't allow the system to adjust.

The second mistake is failure to detect, that is not making a change or a decision on a specific process because you think the variance is OK, when it is actually a cause that can be isolated and prevented.

In both cases, you are working on the wrong problems or failing to detect problems before the customer.

We Culture Behavior:
Identify and Analyze Problems Productively

Any outcome that can be measured, such as the commute time to work or school, will show fluctuations from one day to the next. Deming would say, "Variation is life or life is variation."[88]

How hot your coffee is during breakfast or how long a commute to work takes varies depending on the weather conditions, traffic, or mood.

Key Insight

Variation is expected in every process. What matters is to identify if we need to react to the variation or not.

If you worry about any slight variation in your day-to-day processes, it's hard to delegate decision-making. However, there are ways to identify normal causes of variation (no further action needed) from causes that need our attention. By defining these two categories beforehand for every process, employees are better equipped to identify actual risks and actions needed.

Every problem has common and special causes (see Figure C6.4). Team members need to learn to analyze if what has caused the problem is a common or a special cause, to identify if there is an action that needs to be taken. This is an important consideration to avoid overreacting and stressing over minor issues or overlooking issues requiring attention.

Figure C6.4   Two sources of variation: common and special causes.

The control chart is a statistical tool used to help decide when a process is under control (only common causes of variation) or out of control (special causes appear) and requires attention, just by looking at the data points (see Figure C6.5).

When employees learn to interpret their own processes and variations through simple statistical methods like control charts, they can make better and faster decisions, improve their self-confidence, and provide valuable insights to help their teams.

Basically, you draw a control chart by collecting data while you work. For instance, you can measure the duration of your meetings.

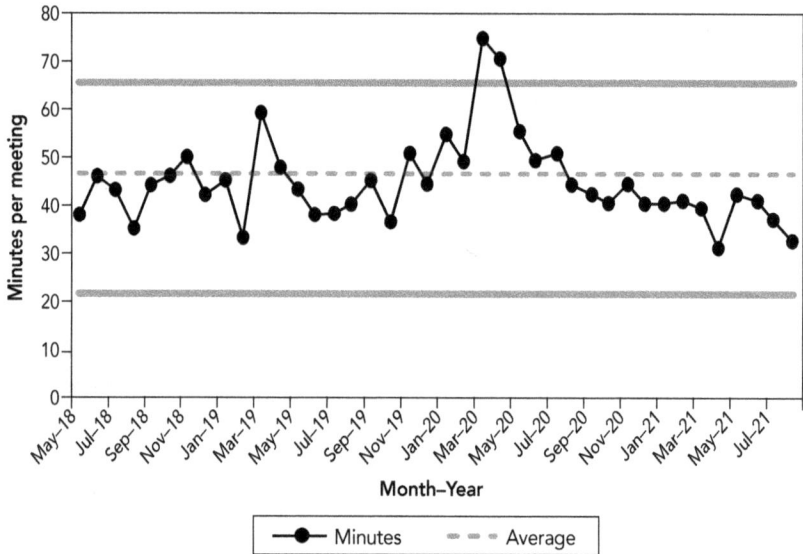

Figure C6.5   Control chart.

This chart shows that the average duration of meetings is typically around 47 minutes (center line). Still, in this sample, if you calculate the average, it is 57 (the sum of the average duration per month divided by the number of months on the list).

March 2020: 75 minutes

April 2020: 70 minutes

May 2020: 55 minutes

June 2020: 49 minutes

July 2020: 51 minutes

August 2020: 44 minutes

Is this good or bad? Well, that depends on what you want. You can see that meetings during March and April were too long, way over the upper limit. And you can also see from the graph that after September, there is a decreasing trend—all meetings were below the average, between 44 and 31 minutes. More than seven points below the average means there was a shift in the process of doing meetings; maybe the organizer changed, or the agenda was scheduled differently.

Is there anything to be worried about? Data points between the red limits (upper and lower control limits) seem to be OK; there are only two that need attention. You need to determine why it took so long (March and April 2020) and then look for ways to prevent that from happening in the future. From what the chart can tell, these two are special causes. Ask your co-workers what happened during those meetings, what went differently. It was likely related to the COVID-19 pandemic surge.

Do you want to intentionally move the average down to have shorter meetings? Work on the common causes. That is, there was nothing wrong with the meetings before March 2020, but in a remote work environment, you may want to have shorter meetings because people lose focus faster. Maybe you would like an average of less than 47 minutes. You will have to change the meeting process: modify the agenda, invite fewer people, set a timer, etc.

Is the shift in the process wrong? It may mean that finally meetings are more effective and organized. On the other hand, maybe you are discussing fewer topics. Maybe the organizer is not allowing other team members to propose ideas, or invitees are not as engaged, reducing the total time of the meeting. What is right for you at this time? Ask your teammates and find out.

You should now understand that numbers don't tell the whole story. Unless you know why there was a shift, you need to determine if the shift is going in the right direction (based on quality, what the internal customers or invitees want for the meeting) and incorporate the lessons learned.

*Download the We Culture app to find out more
about control charts and causes of variation
at www.theweculture.com.*

Incorporate Lessons Learned to Improve Knowledge Sharing

We can have the most robust process, but data points will vary. And no matter how much you analyze the issues, you still can make mistakes. Now that you understand that making mistakes is usually part of every process, you should incorporate a method to learn from previous errors on top of risk analysis.

In November 2019, right before Larry Page and Sergey Brin announced they were stepping down from their management roles at Google, I visited Google's headquarters in Mountain View, California. It was an extraordinary experience to get to know the Google complex from the inside out, as the first time I tried to visit I could only go around the exterior. The restaurants inside the building, the offices that look like vacation centers, the autonomous lawnmower, and many more perks that looked awesome at that time, may now seem totally worthless with hybrid teams. Change is a constant, even for big corporations. And so it is learning.

I also learned a lot about Google's culture from that trip, things that may not be found in books. One of the takeaways was the importance of knowledge management, especially in knowledge-based and project-oriented organizations, and the role of lessons learned. Lessons learned are the knowledge and understanding gained by experience. It is a process of capturing information regarding successful and unfavorable events, and analyzing and transforming that information into knowledge to prevent something from happening again or to promote it as a best practice.

A lesson learned must be:

- Significant: have a measurable impact; can be both successful and unfavorable events
- Valid: factually and technically correct
- Applicable: identifies a specific design, process, or decision that reduces or eliminates the potential for failure, reinforces a positive result, or provides data not opinion
- Able to ask three fundamental questions: what happened, why, and how
- A means to emphasize knowledge sharing: applying the learning locally within the project site or sharing globally or across projects and sites

What to Consider to Build Successful Lessons Learned Process

The company's culture should promote that the quality of the lessons learned is far more critical than the number of lessons learned. Particularly, it should reinforce psychological safety to ensure that even the lessons we want to forget about are shared.

Google implements closed-loop learning (CLL) as a process of applying the acquired knowledge at appropriate life-cycle phases to prevent risks or repeat success. Events that significantly impact cost, schedule, and quality require a CLL session.

Lessons Learned and Retrospectives in Agile Cultures

Agile cultures that implement methods like Scrum have already incorporated lessons learned systems at the end of their "sprint" called the retrospective. According to the Scrum guide[89] published by Ken Schwaber and Jeff Sutherland, the sprint retrospective is an opportunity for the Scrum team to inspect itself and create a plan for improvements to be enacted during the next sprint.

David Horowitz, CEO and co-founder of Retrium,[90] a retrospective software, says "Retrospectives are a chance for your team to inspect how it works and to adapt going forward."

The retrospective is a meeting at the end of each sprint (in Scrum, projects are divided into smaller modules called sprints with a specific length, such as two weeks or one month) that aims to find the lessons learned of the sprint: how it went with regard to people, relationships, process, and tools; identify and prioritize the major items that went well and potential improvements; and create a plan for implementing improvements in the upcoming modules and projects.

In most cases, a Scrum master or an agile coach will facilitate retrospectives to ensure everyone on the team feels safe to collaborate. Most retrospectives focus on discovering answers to three questions: what is working well, what is not working well, and what should be changed.

A *postmortem* is another process, right after a project is finalized, to determine which parts of the project were successful or unsuccessful. In his book *Creativity Inc*,[91] Ed Catmull describes how Pixar used postmortems after a film to ask the team involved to come up with five things they would do again and five things they would not do again.

When many employees leave a company, rotation is high, or even when teams are distributed, the knowledge must be maintained intentionally. The challenge is collecting and disseminating knowledge to benefit those inside and outside of the project. Companies should start incorporating a lessons learned process through periodic closed-loop learning sessions; retrospectives; postmortems; strength, weakness, opportunity, threat (SWOT) analyses; or any other tool that ensures lessons learned are analyzed, shared, and implemented.

Normalize Errors

Brené Brown, in *Daring Greatly*,[92] recommends four strategies for building shame-resilient organizations:

1. Support the leaders who are willing to facilitate honest conversations and help to find potential in people, not those who just prefer to blame or make people feel shame. Employees with shame pass it on to their customers, co-workers, and families. Leaders are the first ones who don't need to start a hunt because of an "audit finding" but need to start a conversation. They need to accept that they are far from perfect, like everyone else, and that is OK.

2. Facilitate an effort to recognize where the organization is using blame or fear to solve problems. Innovation, learning, and creativity are not certain; therefore, they are connected to the fear of failure. If failure is not embraced by the company culture or if it's punished, someone must be found responsible when there is something wrong. So instead of focusing on solving the problem, the focus is on who is to blame. Here there is no room for innovation, learning, and problem-solving. Therefore, disengagement and poor performance are the first symptoms. People stop caring because they feel hurt and vulnerable, wondering if they will be the next person who fails.

3. Normalize errors, fear, and failure by talking about them and bringing them up during one-on-one sessions, courses, or coaching sessions. Identify what you know, admit you have a problem, and make a list of things you

don't know or need to fix. Finally, engage your team in finding the right problems and the right opportunities for improvement.

4. Train all employees on the difference between guilt (you did something bad) and shame (you are bad). Shame causes disengagement, but guilt is more positive, as you have the feeling that you can still learn and improve. Everyone should be trained on how to give and receive feedback so it is more constructive than destructive.

We Culture Tool to Increase Attention: RESET

Sometimes our companies—or worse—our own habits, drive us to stay in the NEA state (negative emotional attractor), that is, in the negative, fear, and number-focus approach. This disconnection hurts and means real pain for us, even when it may only be in our heads.

Managed correctly, stress can drive peak performance and the ability to stay in the zone of flow. When the anxiety passes, the heightened state is followed by a state of recovery and rest. But if the stress is prolonged (such as in the case of the 2020 Summer Olympics, the uncertainty was prolonged for a year), it can lead to chronic stress,[93] which is associated with mood swings, reduced empathy, and impulse control. Gallup research[94] revealed that it affects about 67% of full-time workers at any given time.

This happens to workers as well as in other areas of life. It happened publicly to gymnast Simone Biles when she withdrew from a competition in the 2020 Summer Olympics in Tokyo. So much pressure jeopardized her ability to focus on her flips and jumps. You can talk about it to try to release the pressure, but some cultures still don't welcome this openness and stigmatize those for disclosing it.

You may also miss opportunities or competitor moves because you are too focused. Life becomes a chore, a bunch of goals that you need to achieve, something you ought to do. You don't enjoy it. Then you miss a lot about your life and the lives of others around you. This is when you get "disconnected." Many people suffered this disconnection during COVID-19 pandemic. Many of the distractions they had in their day-to-day lives disappeared, so they turned to increase other distractions, such as Slack, Messenger, more meetings, and more emails. A Microsoft[95] research showed that between February 2020 and February 2021:

- Time spent in Microsoft Teams meetings more than doubled (2.5 times) globally.

- The average Teams meeting was 10 minutes longer, up from 35 to 45 minutes year-over-year.

- The average Teams user was sending 45% more chats per week and 42% more chats per person after hours.

You can do something to break the negative loop and move the stage of emotional engagement and respect for yourself and the people around you by changing your own behaviors.

Daniel Goleman, in his book *Altered Traits,*[96] famous for his work on emotional intelligence, also discusses the importance of attention in companies and in personal life. He says, "most every kind of meditation entails retraining attention."

> *Attention fatigue in radar operators was the practical reason this very aspect of attention had been intensively researched during World War II, when psychologists were asked how to keep operators alert. Only then did attention come under scientific study. Ordinarily, we notice something unusual just long enough to be sure it poses no threat or simply to categorize it. Then habituation conserves brain energy by paying no attention to that thing once we know it's safe or familiar. One downside of this brain dynamic: we habituate to anything familiar—the pictures on our walls, the same dish night after night, even, perhaps, our loved ones. Habituation makes life manageable but a bit dull.*

> *By zooming in on details of sights, sounds, tastes, and sensations that we otherwise would habituate to, our mindfulness transformed the familiar and habitual into the fresh and intriguing. This attention training, we saw, might well enrich our lives, giving us the choice to reverse habituation by focusing us on a deeply textured here and now, making "the old new again.*

The way we live today, with constant distractions, apps, and priorities asking for our attention, it is difficult to focus. We have attention fatigue too. So, we tend to use metrics to simplify our life. But as described previously, we habituate to a way of life that we don't enjoy and prevent ourselves from growing and helping others grow. This is why many companies are starting to offer mindfulness sessions, one-on-one coaching, listening skills training, and more training on observing trends and big data.

You need to balance the positive and negative thoughts – this is what is called a renewal or a *reset*.

Five Ways to Reset and Reduce Stress at Work

It's not only about the things in your life that you can't control but also how you deal with them. Managing stress at work is ultimately about managing yourself, becoming a hero in every area of your life.

You can train your mind and body to react in a certain way that is more productive and satisfying for you. Amishi Jha, in her research at Miami University,[97] shows that we can train to bias our brains the way we need to pay attention to what we want more often.

Over time, the better you manage different situations, the better you will become in the next situation. Eventually, you will master the skills of attention, though it will not be easy. You will begin to manage stress not at the time of the stress but way before it happens.

Remember, it's all about your routine. So let's talk about the habits that you want to incorporate into your daily life and within your team to improve performance and experience more joy at work.

- Relax: Calm your mind listening to your breath.
- Enjoy: See the bright side in everything.
- Simplify: Only keep what you need.
- Exercise: Turn thoughts into actions.
- Thank: Appreciate what you have.

Relax: Calm Your Mind Listening to Your Breath

Stress activates the body's fight-or-flight response, so blood pressure and heart rate go up. Hormones like epinephrine and cortisol are released into the bloodstream. In the short term, memory and creativity are improved. However, once stress exceeds a certain threshold, then performance decreases. If you have to speak in public, you become out of breath. If you have to make a quick decision, your heart rate becomes too high to let you think; You become so stressed that you don't even realize you are under stress, so you cannot think of how to overcome it.

How do you overcome it? You have to establish a daily routine when you invoke a state of profound rest by simply breathing deeply. You can do meditation, yoga, a nature walk, or simply sit on your bed and breath consciously, relaxing your muscles progressively. Amishi Jha says that 12 minutes of mindfulness a day, for instance, can help improve attention. Take time to listen to your breath. This helps you calm your mind and gain control. Learning a relaxation response is not instant, it takes some time, which is why you need to start practicing before you need it. The good thing is that you can do it anywhere, anytime, at no cost! Every time you feel exhausted or under pressure, call your relaxation response by breathing deeply.

Enjoy: See the Bright Side in Everything and Celebrate It

Once you are under the control of your body, you are better able to choose your thoughts; they can be positive or negative. Examples of negative thoughts are when you fear an unexpected outcome or imagine that you will be punished or ghosted (someone cuts off all communication without explanation).

Help yourself and your team start and finish the day seeing yourself doing a good job. Avoid thinking only about what went wrong or having negative expectations about the future. Avoid counterproductive thinking that is a constant rumination in your mind about what went or could go wrong. Notice the present, what has been done, and enjoy it and celebrate it.

Every time you finish an important task, such as a presentation to a client, think about what went wrong and write down how to correct it in the future. But also think about what went well (spend more time on this). Be balanced on how you criticize yourself. If you think about it, there is a good side and bad side to everything—it's all up to you and how you want to see it. The Netflix movie *About Time* captures this thinking wonderfully.

About Time—Happiness scene
https://youtu.be/YVm8NnUzbXk

Simplify: Only Keep What You Need

For sure the main excuse at this point will be that you don't have time to breathe, relax, or enjoy. You are still in the numbers-focused mindset; breathing doesn't make money. So how do you make up for more time? You simplify your work life. Check your desk, your office, your wallet, your purse, your mailbox! How many things do you see that you don't need? How many of them bother you or distract you? How many of them keep you busy in exchange for nothing really rewarding? How many meetings do you attend or tasks do you perform that add no value to you or your team? Think about it. No matter how small, every minute counts.

- Reduce if you have too many.
- Eliminate if you don't need it.
- Move out if it may be useful somewhere else.

Review your daily routine and eliminate all the waste you see and do every day, focusing only on what matters. Having fewer things to think about helps you focus and prevents your brain's burnout process.

I always use the 5S methodology to help me with this. You will find out more about it in Skill #12 or by reading my book *5S Your Life*.[98]

Exercise: Turn Thoughts Into Actions

The fourth habit is about turning thoughts into actions. Sometimes it's hard for us to manage our thoughts, they just come and go, even when we are trying to breathe and focus. Turning thoughts into actions by writing them down is a wonderful tool.

When you have to concentrate on your body movements, you don't let your problems affect your emotions. As the Mayo Clinic[99] explains, it's like meditation in motion. It helps you exchange new oxygen and pump up endorphins that help to improve your mood. You'll often forget the day's challenges and concentrate only on your body's movements. In the past, companies wouldn't allow employees to exercise during work hours, but now it's becoming an accepted practice, and it is even promoted by HR. In hybrid workplaces, when you don't even have to walk out the door to go to work, doing some exercise should be part of your schedule.

If there is a problem that blurs your mind, gather with your team, virtually or face-to-face, and share all the information you have in a brainstorming session. Find solutions as a team by doing the exercise of writing down what you have in mind, organizing the ideas, and prioritizing. If ideas don't flow yet, an outdoor activity can also help refresh the neurons. If you are at home, you can go out and exercise (for at least 10 minutes!). Engage in an activity that calms or distracts your mind, such as walking, running, or even cooking. I love skating and swimming. Organizing or doing 5S also works, as I call it "meditation in action." The more challenging the activity, the better. Some people love climbing, for instance.

Thank: Appreciate What You Have

Stress is the reaction we have to excessive demands placed upon us. It is usually triggered when we are unsure if we can cope with the problem or we fear the outcome.

Make sure you establish a routine to review projects in various stages to avoid surprises at the end, and thank your team for the effort in public at every stage. Remember that if the outcome is bad, it is probably partly your fault as well. Maybe the goals were not clear enough or the team doesn't have enough resources. So share both the bad and good news, but especially share the rewards. Rewards come in different forms: smiles, thank you notes, certificates, days off, you name it. As a leader, the reward I cherish the most is when a team member enjoys the project and is ready to start the next one. Practice together some "I am grateful about" routines.

Just before going to bed, include in your routine a moment to thank (your god or the world in general) for what you have received during the day. It could be a work accomplishment or a simple hug from your daughter; just try to think about anything that made you feel happy during the day. Thank yourself, eyes closed, and breathe deeply, relaxing your muscles. If you don't have anything to thank for, you are too hard on yourself. Think again. You may get ideas during that session. Write them down. Keep a journal next to your bed. Write down all the ideas that you have in mind. Spend more time thinking about what went fine and finish the session with a smile on your face. You will see how different you will feel the following day.

Follow the same night routine every day. Just like babies love routines, you should love them too, at least to accomplish the things from which you can't escape.

Always do it at the same time to enhance the sense of ritual and establish a habit. If you practice it every day, it will be easier over time, so you will be able to enjoy the routine and your other tasks to accomplish. If you are not enjoying it, go back to square one!

Hands-on 6.1

Analyze your work process and describe the different functions you and others perform. Identify failure modes (what could go wrong), and determine the effects of the failure, why it could go wrong, and what you can do to solve or prevent it.

Exercise

Describe the functions in your process and determine what could go wrong.

Identify FAILURE MODES. What could go wrong?	Identify FAILURE EFFECTS. What happens if it goes wrong?	Why it could go wrong? How can you solve it?

Summary

Achieve quality by thinking and deciding together.

Recap 1
Push decision-making to the front line.

Recap 2
Understand and share the why behind goals and performance metrics.

Recap 3
Manage risk by sharing, observing, and listening.

Reflection Time

Take five minutes to think about three highlights from the dimension of ATTENTION. Write them on your note pad or the action plan available on the We Culture app.

Three Highlights

_____

_____

_____

References

80. Ed Catmull with Amy Wallace, *Creativity Inc.* (New York: Random House, 2014).

81. Eric Ries, *The Lean Startup: How Today's Entrepreneurs Use Continuous Innovation to Create Radically Successful Businesses* (New York: Crown Business, 2001).

82. Agile Alliance, "Manifesto for Agile Software Development," https://www.agilealliance.org/agile101/the-agile-manifesto/.

83. Jeff Dyer, Hal Gregrersen, and Clayton Christensen, *The Innovator's DNA: Mastering the Five Skills of Disruptive Innovators* (Brighton, MA: Harvard Business Review Press, 2011).

84. Brené Brown, *Daring Greatly: How the Courage to Be Vulnerable Transforms the Way We Live* (New York: Penguin Random House, 2012).

85. Brené Brown, *Daring Greatly: How the Courage to Be Vulnerable Transforms the Way We Live* (New York: Penguin Random House, 2012).

86. Frederic Laloux, *Reinventing Organizations*, (Brussels, Belgium; Nelson Parker, 2014).

87. W. Edwards Deming, *The New Economics for industry, Government, Education,* third edition (Cambridge, MA: Massachusetts Institute of Technology, 2018).

88. Mustafa Shraim, Deming, https://deming.org/impact-of-process-tampering-on-variation-experiment-presented-at-engineering-education-conference/.

89. Ken Schwaber and Jeff Sutherland, "The Scrum Guide, The Definitive Guide to Scrum: The Rules of the Game," *Scrum Guides,* https://scrumguides.org/download.html.

90. Retrium, "What is a Retrospective," https://www.rctrium.com/ultimate-guide-to-agile-retrospectives/retrospectives-101.

91. Ed Catmull with Amy Wallace, *Creativity Inc.*, (New York; Random House, 2014).

92. Brené Brown, *Daring Greatly: How the Courage to Be Vulnerable Transforms the Way We Live* (New York: Penguin Random House, 2012).

93. Jan Ascher and Fleur Tonies, "How to turn everyday stress into 'optimal stress,'" *McKinsey & Company*, 2021. https://www.mckinsey.com/business-functions/people-and-organizational-performance/our-insights/how-to-turn-everyday-stress-into-optimal-stress (accessed February 2021).

94. Ben Wigert and Sangeeta Agrawal, "Employee Burnout, Part 1: The 5 Main Causes," *Gallup*, 2018. https://www.gallup.com/workplace/237059/employee-burnout-part-main-causes.aspx.

95. Microsoft, "2021 Work Trend Index: Annual Report, The Next Great Disruption Is Hybrid Work — Are We Ready?" *Microsoft*, March 2021.

96. Daniel Goleman and Richard J. Davidson, *Altered Traits: Science reveals how meditation changes your mind, brain and body* (New York: Avery, 2017).

97. Amishi P. Jha, *Peak Mind: Find your focus, own your attention, invest 12 minutes a day* (HarperCollins Publishers, 2021).

98. Luciana Paulise, *5S Your Life: Stop Procrastination and Start Self-Organization*, 2020.

99. https://www.mayoclinic.org/healthy-lifestyle/stress-management/in-depth/exercise-and-stress/art-20044469.

Part 3:
RESPECT

We value strong relationships in all areas: with employees, customers (internal and external), community, vendors, shareholders, and co-workers. Strong, positive relationships that are open and honest are a big part of what differentiates Zappos from most other companies. Strong relationships allow us to accomplish much more than we would be able to otherwise. A key ingredient in strong relationships is to develop emotional connections.[100]

T he third dimension to develop is our "emotional" side: respect for everything and everyone, from respecting every team member, client, or supplier we interact with, to respecting the established rules, processes, and assets.

Respect, according to Merriam-Webster's dictionary,[101] means "to consider worthy of high regard." In regard to people, it means that you accept them for who they are, even when they're different from you or you disagree with them. Respect in your relationships builds feelings of trust, safety, and well-being. Being respected in the workplace is one of the things that matters most to employees. They especially want leaders to respect them.

The emotional side of people and organizations is not usually considered as important as the rest of the intelligence or dimensions. Productivity or results seem to be more important than engagement. The HR department is often seen as a merely administrative area. Research shows that most CEOs undervalue their chief human resources officer (CHRO).[102]

Still, multiple researchers, starting from Mayo to Deming and Daniel Goleman to Amy Edmondson, have shown the importance of handling emotions for success. Emotional intelligence has become a key skill set throughout our lives, whether leading a team, raising a family, or dealing with your neighbors. Leaders in all industries struggle with emotional intelligence, starting with Tony Hsieh, former CEO of Zappos, which triggered his internal need to build a company that would "deliver happiness."

Fortunately, after the COVID-19 pandemic and the constant urge to embrace diversity, the dimension of respect is no longer overlooked. There are other functions that work hand-in-hand with the CHRO, such as the people officer or culture officer. All should work with the CFOs and CEOs to reshape the organization the agile way.

How Can You Build a Work Environment Where Respect Reigns?

As you have been reading so far, in a we culture, everyone is considered important. Hierarchies don't really exist, as no position is more important than the others, from owners and managers to line employees, contractors, suppliers, customers, maintenance, and administrators. Each person who is part of the company's process is vital to achieving the results.

Teams need to be agile to adapt to volatility, uncertainty, complexity, and ambiguity (VUCA) and require people who bring different skills, backgrounds, and experiences to the table. Having a variety of knowledge in just one team increases the potential for finding innovative ideas to solve problems, increases autonomy, and therefore, increases the agility to react.

To build an environment where respect reigns, the first and easiest way to start is by respecting the rules, processes, and assets of the company. This is also part of the We Culture. Rules and processes are agreed upon and written by team members to follow and standardized as best practices. If they are not respected, the customer may be impacted, receiving a defective product. Another internal customer may be impacted too, by redoing what was been done before by previous co-workers.

For example, employees know that they need to organize the tools they used before leaving their workstation. But at some point in time, they may be in a hurry; for instance, their family may be waiting for them, so they don't have time to organize the tools for the next shift. Then, other workers in the next shift arrive and find all the tools out of place and dirty. They get frustrated and decide they don't have to organize the tools before leaving either. A negative balancing feedback starts. That's how respect works when an agreement is broken. Employees feel frustrated when others don't follow the rules, so they become disrespectful, too. The imitation game corrodes camaraderie and engagement.

Respect is a crucial aspect of an organization to drive contagious positive behaviors that engage all collaborators to work on the same side.

The essential behaviors to develop a culture of respect with the company that will increase employee engagement are valuing and increasing diversity, equity (also referred to as equality), and inclusion throughout the employee experience; promoting psychological safety at the team level; and opening intimate one-on-one conversations at the individual level to get to know and empathize with each individual. The communication process increases engagement when respect throughout the employee experience is the input (see Figure P3.1).

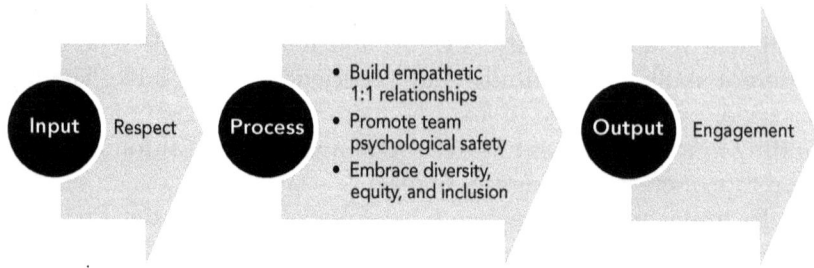

Figure P3.1   The process of communication in the respect dimension.

Respect is the input for these three desired behaviors that will drive higher engagement:

- Build empathetic relationships, one person at a time

- Promote team psychological safety

- Embrace diversity, equity, and inclusion

References

100. Zappos, "Zappos 10 Core Values," *Zappos Insights*, https://www.zapposinsights.com/about/core-values.

101. https://www.merriam-webster.com/dictionary/respect.

102. Ram Charan, Dominic Barton, and Dennis Carey, "People Before Strategy: A New Role for the CHRO," *Harvard Business Review*, 2015, https://hbr.org/2015/07/people-before-strategy-a-new-role-for-the-chro.

7

Skill #7: Build Empathetic One-On-One Relationships

In the Project Oxygen study, Google found that higher-scoring managers are more likely than lower-scoring managers to have frequent one-on-one meetings with their team members. Meeting frequently and individually with team members can require a large time investment but can identify issues early and provide a forum for the manager to give feedback and guidance.[103]

B uilding one-on-one relationships is key to productive feedback. Giving, receiving, and asking for feedback is rare in most companies. It terrifies both the employee and leader. Still, doing it right is one of the drivers of high psychological safety, equality, and inclusion.

The Great Resignation

Companies are learning after the COVID-19 pandemic crisis that the one-size-fits-all practices don't work anymore. Employees are demanding to be respected the way they are, to be considered as whole individuals. One may prefer working from home, while another loves coming to the office every day and staying long hours.

Still, few companies allow employees to choose, for instance, when or where to work. Many companies allow employees to access the office at any time, even at night or on weekends. But their rule is: Yes, you can go and work at night, or you can work from home, but I still want you at your desk from 8 a.m. to 5 p.m.

According to the Labor Department, once Americans started getting back to the office after the main pandemic crisis in April 2021,[104] a record 4.2 million people quit their jobs. In September 2021, a record 4.4 million people quit. People began to see their lives differently. They realized how much time they were spending commuting when they could be doing the same work without leaving their homes. Others noticed how close they had become to their kids and felt they would lose that connection again. Some had been traveling worldwide while working remotely and didn't want to get back to a pricey apartment in New York City or the Bay Area. This period was called the "Great Resignation," when employees started leaving their jobs, and searching for more money, greater flexibility, and greater personal contentment.

More genuine interactions with co-workers will help foster a workplace where people feel more comfortable being themselves.

A Great Team Needs Empathy

Companies cannot create rules that apply to or satisfy everyone. But they can trust their leaders and team members to accommodate the rules without breaking them to ensure they fit more tightly and bring the flexibility employees are craving.

A way to adapt the rules to the specific needs of each team member is to open intimate, one-on-one conversations to get to know and empathize with each individual.

Empathy is now more critical than ever, given that people are experiencing multiple kinds of stress and a decline in mental health. A global study by Qualtrics[105] found that 67% of people are experiencing increases in stress, while 57% have increased anxiety, and 54% are emotionally exhausted, 53% are sad, and 50% are irritable. This study also found that when leaders were perceived as being more empathetic, people reported greater levels of mental health. Eighty-eight percent of employees and HR professionals would be more likely to stay with an employer that empathized with their needs.[106]

Empathy can contribute to positive experiences for individuals and teams. A study of 900 employees by Catalyst,[107] a nonprofit organization that helps build workplaces that work for women, found empathy has some significant effects on performance:

- **Innovation:** 61% vs 13% of people with highly empathic senior leaders report often or always being innovative at work compared to people with less empathic senior leaders.

- **Engagement:** 76% vs 32% of people with highly empathic senior leaders report often or always being engaged, compared to people with less empathic senior leaders.

- **Retention:** 57% vs 14% of white women and 62% vs 30% of women of color said they were unlikely to think of leaving their companies when they felt their life circumstances were respected and valued by their companies compared to those who didn't feel that level of value or respect for their life circumstances.

- **Inclusivity:** 50% vs 17% of people with empathetic leaders reported their workplace was inclusive, compared to those with less empathetic leadership.

- **Work-Life:** 86% vs 60% of people felt that when their leaders were more empathetic, they were able to navigate the demands of their work and life — successfully juggling their personal, family, and work obligations compared to those who perceived less empathy.

- **Burnout:** 54% vs 67% of employees with highly empathic senior leaders were less likely to report high levels of general workplace burnout than those with less empathic senior leaders.

Senior leaders can establish an environment of empathy within the organization, but immediate leaders play a key role in developing one-on-one relationships with the employees.

What is Empathy Anyway?

Empathy means putting on the shoes of the other person and trying to see with their glasses. For example, when you create a product, you want to know how the customer feels, not how you feel about it, to help solve their problem. Is it easy to use? Do they have any impediments to using it? Do they use it in a different way than others? Do they have to change some of their behaviors to accommodate for the product to be more helpful? It is the same with team members. You have to find out what they need to succeed.

The challenge is that most employees won't openly discuss what makes them unique or what different needs they have. Companies have been communicating for years since Taylorism and Fordism that if you are or act different, you will suffer the consequences of being left out of groups or experiencing less growth.

Now it's time for leaders to prioritize respect and appreciate individual differences. The leader's job is not to cheer up the team or tell them what to do. Leaders don't need to be experts in psychology or mental health either. Instead, leaders just need to be available and check in with their team members to build trust, provide support, and offer flexible options that accommodate different needs. Leaders need to build relationships at the one-on-one level as a regular practice and an essential part of their work. This is their way to empathize with their team members and learn to value their differences, highlight their strengths, and help them trust their abilities.

Some behaviors that help leaders be more empathetic are based on the Catalyst survey:

- Start meetings with personal check-ins.
- Pay attention to what employees do, including their body language, not just their words.
- Ask about family.
- Start with a sign of gratitude.
- Clearly say when they are doing a good job.
- Say things like "I am sorry."

Me Culture Routines: Annual Performance Reviews

Gallup research[108] asked in a survey, "How often do you receive feedback from your manager?" The replies were:

Daily: 7%

A few times a week: 19%

A few times a month: 27%

A few times a year: 28%

Once a year or less: 19%

Only 26% of employees strongly agree that the feedback they receive helps them do better work.

Many companies have a process of providing feedback to the employee on an annual or semi-annual basis. Even when some conversations may be held throughout the year, this one annual instance is the most important because they usually communicate salary increases, promotions, or improvement opportunities. Unfortunately, these sporadic feedback sessions are used to summarize events that happened in the past, in the last 6 or 12 months, and build a story to justify how they will be treated next year. These performance reviews barely help employees improve, as they normally assign a ranking to employees compared to others, not considering their own growth.

Usually, the message is: you are outstanding, good, or need improvement. If the performance is "needs improvement," they are on the verge of dismissal. Predefined quotas say that every group should have at least 5% of the personnel in this category (no matter how good they are, and sometimes the percentage is even higher). You can't have 10 excellent performers out of 10. You always need to have an outstanding performer, eight good ones, and one needing improvement. A good performer could rank low on an outstanding team, while that same performer could rank high on a lower-performing team. Does this seem fair enough?

The problem is that whatever the employee did right or wrong, it was months ago (if the review is annual or semi-annual). It frustrates employees because there is little they can do to change it now. The feedback is too late. If they need improvement, they may have a grace period of three months where they have to be pushed to excel or leave. If the performance was satisfactory, they didn't have the chance to get even better or replicate it. They know just now. Unfortunately, they didn't get the praise at the right time.

In the end, there is no two-way communication; employees are just told how they performed and what should change in the future. This frustration correlates, as mentioned before, with increased levels of anxiety, depression, and disengagement, with a tremendous negative impact on performance right before and after the evaluation. Nine out of ten employees feel they are not good enough; they feel inferior.

We Culture Behavior: Provide Feedback

Me routines like annual performance reviews are used to provide feedback for the individual, but what they mostly do is erode the employees' psychological safety systemically. Leaders could avoid their negative impact simply by improving one-on-one conversations and making them more empathetic.

Individual reviews should be done more often and closer to the particular event that needs to be highlighted. Performance is not based on one single activity but a series of activities, results, and behaviors that can hardly be evaluated in just a one-hour session. Every activity should be discussed separately at the time it is performed, in as much detail as needed.

Feedback needs to be part of their day-to-day, to help employees improve on a daily basis, not an unusual event that creates unease. Ideal feedback sessions should look more like conversations in which two adults express their views and concerns, questions arise, and both agree on what can be done.

Leaders no longer need to look like heroes who have *all* the answers to be successful. Organizations are too complex, which is why leaders have to help the team become the hero, not the other way around. Leaders have to facilitate interactions and become coaches.

Individual reviews should be set not for criticism but for discussing ways to continue growing. And they should not be used to communicate one-way a diminishing ranking. What is the goal of telling a person he or she did not do well enough if you can't tell that person how to improve or he or she can't even discuss it? Unless the outstanding person leaves the group, the good ones cannot perform better. That doesn't make any sense. The best feedback is the kind that makes you feel proud of your strengths and helps you grow on your improvement opportunities, not the kind that pushes you to be better than others or peril. Basically, reviews should help people work better for the team, not for their own benefit.

Feeling valued is a condition for well-being, while the effect of not mattering if you receive a bad review can be devastating and drive anxiety.

Some solutions Deming suggested in his manuscripts in the 1980s:

"The annual performance review may be abolished." Suggestions follow:

- Hold a long interview with every employee, three, or four hours.

- "Figures on performance should be used not to rank the people in a group that fall within the system, but to assist the leader in accomplishing improvement of the system. These figures may also point out to him some of his own weaknesses."

- The formula for raises in pay may involve seniority. However, it will not depend on rank within the group.

- Use more careful selection of people in the first place.

- Provide better training and education after selection.

- A leader, instead of being a judge, will be a colleague, counseling and leading their people on a day-to-day basis.

- A leader will discover if any of their people are outside the system using statistical data.

Key Insight

Leaders no longer need to look like heroes who have all the answers to succeed. Organizations are too complex. Leaders have to facilitate interactions and become coaches.

We Culture behavior: New ways to provide feedback

Understanding employee challenges and concerns is vital. That's why leaders should organize periodic meetings to check in one-on-one, especially in hybrid teams and in times of uncertainty. When working face-to-face, a leader can determine how an employee feels by looking him or her in the eye or paying attention to new behaviors. In remote work, though, employees' feelings can be harder to perceive.

Therefore, coaching sessions should be even more often than when working in the same office. Leaders have to build an environment where they can ask questions, offer continuous feedback, and receive questions and concerns. Remember that employees are closer to the customer than leaders are, so feedback can be about themselves or their work and the process and the customer.

Gallup research[109] shows that only 14% of employees strongly agree that the performance reviews they receive inspire them to improve. It's key to train leaders to provide meaningful and timely feedback, especially if some team members are working remotely.

Now, let's look at different ways to provide feedback. Some organizations, usually the ones applying agile or self-organizing methods instead of the annual or semi-annual evaluations, promote constant feedback between peers. Some do not have any formal performance evaluation, such as the European company FAVI.[110] Other companies simply have informal conversations once a year with their work team to focus on strengths and opportunities for improvement. In Buurtzborg[111] (a nonprofit organization for nurses from the Netherlands with more than 7,000 employees), employees are put into groups of three people, each one prepares their own evaluation plus the assessment of their two partners, and they share them.

Agile organizations that implement Scrum methods have as part of their routines what is called a retrospective, which helps team members "inspect themselves" regarding how the team performed in terms of people, relationships, process, and tools (not numbers) after each sprint or milestone. Did they allow everyone to participate? Were there any impediments to teamwork? (see Chapter 11).

Some things to consider:

- A coaching session can be initiated by the coachee or by the coach. If the coachee is the one asking for it, make sure to stay open and available. Don't delay it, just determine the best time and schedule it as soon as possible.

- In case you need to discuss a particular subject referred to a recent event, offer feedback as soon as possible.

- In a one-on-one, you can have different objectives. Depending on the employees' needs, you can be teaching (you provide answers and advice on a specific topic), mentoring (you facilitate growth sharing your experience), or coaching (you help them find their own answer). As a team leader, you should balance the different types of sessions. Just like following the training cycle in martial arts called Shu Ha Ri, first, you teach so the employees can learn. The employees start building muscle memory to change habits by simply following what you teach them. Then Ha is about adapting the teachings to the circumstance. The coachee reflects on the teachings, implements them, and looks for mentoring to experiment new ways. On Ri, you simply coach, they are already knowledgeable, and they can propose the course of action while you provide support and alignment.

- One-on-one conversations are commonly accepted as two-way feedback sessions, but that doesn't mean the feedback is always from the leader to the employee and that the leader needs to propose and guide the session. Employees can, and should, offer feedback to the leader. More than a feedback session, I would argue they should be called *conversations*, which anyone can initiate, at any time, with any objective. They can be planned or simply initiated when there is an opportunity to connect.

- Be open to discussing any issues, even personal matters. If you don't think you can handle it, make sure you guide the employee to find extra help. If you close the conversations to only business-related issues, you may be missing an important story about why the employee is or is not performing or feeling well. A research[112] found that 39% of respondents feel more sense of belonging when their colleagues check in with them personally and professionally. Checking in personally shows you care about your people beyond their work product and even more during a crisis by using phrases such as "How are you handling the___?" And also, be willing to answer the same question in return.

- Start a conversation with a positive approach by using the "appreciative inquiry." This method developed by Dr. David Cooperrider at the Weatherhead School of Management encourages clients to begin by naming and building upon what is going well.

Type of conversations

1. Check-ins: periodic sessions with no defined agenda. Frequency is pretty established, once a week, daily, or monthly.

   a. Focus on the person, uncover hidden needs

   b. Talk about recent events and reinforce specific behaviors (positive or negative)

   c. Review goals, workload, and impediments

   d. Seize small daily opportunities to connect authentically with a simple "How are you doing?" or "How can I support you?" for both business and personal matters

2. Specific need: help the employee find an answer, a solution, or an improvement. These are one-time events with a specific agenda in mind. Frequency as needed (one-time doesn't mean annually, it is just that frequency is not established).

   a. Help solve problems

   b. Assist on specific matters that the employee wants to discuss

   c. Focus on new skill development

   d. Can be pro-active in anticipation of work challenges

   e. Support career development

   f. Mentor the employee when a change is needed

The GAP Model

There are many models used to coach employees. There is the ROPE model (results, opportunities, problems, and execution) and the one used by Google, GROW (goal, what do you want; reality, what's happening now; options, what could you do; will, what you will do?). Find whatever model makes you feel comfortable identifying a goal to accomplish. I call the model I use the GAP model: goal-setting, asking, and planning.

Goal-setting

Define together with your employee the session's goal so you can focus on one topic at a time, determine the urgency, and estimate the duration. With small talk such as "How are you handling X?" or "How are you doing with X project?" you will figure out what might be possible or desired. Identify where you both would like to be after the session. Does the employee need a coach or a mentor? Who will drive the conversation?

Leave room for the employee to propose, ask questions, and express concerns. Empathize with their specific needs. Research published in the *Harvard Business Review* shows that "When employees need your help they are likely experiencing some form of shame, even if it's just mild embarrassment — and the more serious the problem, the deeper the shame. Feeling and expressing empathy is critical to helping the other person defuse their embarrassment and begin thinking creatively about solutions." The easiest way to empathize is to listen actively and leave your own agenda in second place. Remember the Pareto principle (the 80/20 rule): as a leader, speak only 20% of the time, and let your team member talk the other 80% of the time.

Key Insight

The Pareto principle (80/20): As a leader, speak only 20% of the time, and let your team member talk the other 80%.

The goal-setting stage can take almost an entire session if this is the first session of a long-term relationship, or it can take barely 5-10 minutes as a session introduction for an ongoing commitment.

Asking

Drive the conversation by asking questions. This part of the session is the longest, taking approximately 50-60% of the total time. You can ask general questions, or you can ask specific questions that focus on specific actions or reactions contemplated, such as "You seem frustrated with Sean. How's that relationship going?"

Key Insight

Drive the conversation by asking questions.

Look for strengths to accomplish the goal or solve the problem stated in the previous step. Make the employee feel confident about their own skills and identify together any gaps. If you want to help someone change, learn, or grow, let that person figure out how to achieve it. Don't overemphasize results and analytics or what "should" be. That makes people operate in the NEA (negative state, see "Part 2: Attention"), according to Richard Boyatzis,[113] a state of fight or flight, fear, and anxiety. They get more closed to sharing and more focused on being defensive than on interacting positively, and they shut down the capacity to change or learn.

Key Insight

If you want to help someone change, learn, or grow, let them figure out how to achieve it. Do not overemphasize results and analytics or what "should" be. That makes people operate in a negative state.

Teachers, parents, and even doctors tend to push us into an NEA state, saying what is wrong, what our weaknesses are, and what we need to change, providing negative feedback and adding more pressure.

When you focus on the strengths instead, a person's PEA is turned on (positive state), which is essential to learn or adopt new behaviors and actions.

You will notice you are evoking the PEA with questions when the employee's eyes get brighter, and speech gets faster; the employee leans forward to you, is open to possibilities, and feels renewed and curious. When you find a vision, a place where you want to be, it is a tipping point—it is easier to change because you know where to go.

Through questioning, also look for potential issues or threats intentionally. Try to listen actively to understand the concerns and raise your own concerns if you are mentoring. Use phrases like "I can understand why you...." After confirming your understanding, find something to agree with and address their concerns: "Let's see if we can address each concern...." Do not discuss how the employee feels; try to understand why he or she feels that way. Sometimes people get stuck with a block, and by asking questions such as "Have you thought about this?" "Have you tried that?" they may see a different path. By questioning instead of demanding, the coachee gets engaged in the solution and is eager to act.

Key Insight

By questioning instead of demanding, the coachee gets engaged in their own solution and is more eager to act.

Planning

Discuss alternatives to close the gap identified in the previous step. You can propose some options, but primarily let the employee suggest. Like in a brainstorming session, do not judge beforehand. Remember that whenever you try to impose a solution, the coaching has no impact on helping the employee. It only frustrates and weakens the person even more. Closing the gap must empower the person to feel more confident than before and hopeful about the future. It doesn't have to be the perfect answer; the solution can be adjusted and twisted over time. End the conversation with a clear definition of what needs to be done, by whom, when, and how. Finally, agree when you will meet again if needed.

Key Insight

Closing the gap must empower the person to feel more confident than before and hopeful about the future.

Reverse Mentoring

Another alternative is reverse mentoring, which pairs younger employees with executive team members to mentor them on various culture and technology topics. This is a solution to connect, attract, and retain the younger generations that are more prone to leave the organization and become disengaged.

A *Harvard Business Review* research[114] confirms that "Reverse-mentoring programs provide millennials with the transparency and recognition that they're seeking from management." BNY Mellon's Pershing experienced a 96% retention rate for the first cohort of millennial mentors. At Estée Lauder, millennial mentors developed Dreamspace, a knowledge-sharing portal to exchange ideas. PricewaterhouseCoopers launched its reverse-mentoring program as part of its drive for diversity and inclusion. This information can even be used to improve the hiring strategy.

**We culture routines: Inspiring team members
during one-on-one sessions**

- Spend the first five minutes of your one-on-one meetings connecting and observing, that is, just asking questions about that person's life. This is a way to ensure the team member still feels connected emotionally with you. If this is not the case, it will be more challenging to get the employee to listen to you or even change any behaviors. If he or she is not able to look you in the eye, or their answers are short and dismissive, you may want to work harder on identifying the disconnect.

- Practice active listening and ask open-ended questions to facilitate the team member's own insight (questions that start with "what" and "how" encourage expansive thinking). Exercise self-control to withhold your advice.

- Provide specific and timely feedback. Bring to the table specific situations with as much detail as possible and offer feedback as close to the event as possible. Prepare well by focusing on the facts and providing examples. You may need to get input from others involved in the situation.

- Balance positive (motivational) and negative (constructive) feedback and understand the unique strengths and development areas of each team member.

- Be fully present and focused on the team member. Listening actively so you can coach effectively requires eye contact; even if you do it online, try to turn on the camera to detect facial expressions, gestures, etc. Eliminate distractions by turning off your phone, finding a dedicated space where you won't be interrupted, and avoiding multitasking. Avoid interrupting teammates during conversations, as that will establish an interruption rule. Stay focused on listening rather than answering to show respect, and you will receive the same treatment. Demonstrate that you are listening by summarizing what people say after they say it.

- Take notes to remember the most significant findings, but don't try to transcribe everything, or you will get distracted. Just write down words or comments.

- Set clear expectations in the planning and stick to them; otherwise, the team member will miss the point of the coaching session, and it will be harder to commit to a long-term relationship. The team member needs to have the end goal in mind in every session to visualize progress.

- Clarify how progress will be communicated. Does the team member want you to communicate what he or she has accomplished every day? Should you send out an end-of-week digest or wait to have the one-on-ones? Do you expect to meet the goal 100% or just part of it? Objectives and key results (OKRs), for instance, are so challenging that leaders usually don't expect 100% completion. Some leaders value a continuous improvement mindset where a little bit of progress is shown in every session. Clarifying expectations helps the employee avoid frustrations.

- Empathize, don't sympathize. Sympathy means feeling bad for the other person rather than trying to understand him or her. You need to acknowledge the issue but quickly focus on the problem and how to solve it. Feeling bad for team members and letting them talk can be liberating for them, but they also need you to provide a positive outlook and guidance to move on. Focus on what can be done, not on what is lost.

- Never diminish how other people feel. You may think you have bigger problems, or they are overreacting. That is how you see it, but they may not see it the same way. This is very common with DEI issues. For instance, white men could think that a black woman is overreacting to a comment by calling it harassment or discrimination. This could be a microaggression that only the black woman can feel. The same happens with kids; their issues may not seem so significant compared to parents' issues. Still, they need to be heard. Even if you feel like it isn't that serious, listen to them to understand their own perspective and help them find a way out. Sometimes people just need to feel supported, respected, and understood. Sometimes there is a serious bias that other employees need to correct. Addressing the issue is how associates are given more confidence, which will help them find better answers next time.

- Simply believe, don't judge and don't impose opinions.

- Don't simply tell them what to do. When they are part of the solution, they will be keener to implement it.

- Focus on behavior, not the person; do not label the person. Instead of saying things like "you are not communicative" or "you are disrespectful," it is better to say "you did this…" or "this is how they felt about your comment."

- Stay positive about the feedback, not defensive. What others say about you is just their perspective; try to understand why they are saying it by asking questions and examples, the same way you should do with them. Show vulnerability by acknowledging their views and being open about your own challenges. Ask for recommendations or suggestions.

- Assume positive intent: Start any conversation with your colleagues believing that those talking or listening mean well, especially when it comes to difficult issues. It will help you listen more actively and remain positive.

- Plan more carefully when you anticipate a session will be challenging. What are the most important points you want to address? What evidence do you have? How can the other person react? It is important to consider that each one reacts differently, so knowing the personalities of the employees and their personal challenges is decisive. I usually suggest having a short personality test done to team members to help us tailor the message, especially with new team members.

- Inspire them to pursue their dreams and achieve more than they ever thought possible.

How Often?

One-on-one sessions should be held as often as employees require, but stick to a regular calendar if they don't seem to require it.

The number of individual coaching sessions depends on how often you interact with the individual in the everyday operations and how many team meetings you have. If the interaction is constant, like working in an office environment, you may be OK with a monthly meeting unless the employee demands sessions more often.

If working remotely, you may want to do weekly coaching sessions of at least 15 minutes each. In a hybrid work environment, you may stick to doing one-on-ones when the employee is at the office.

Some leaders argue they don't do frequent one-on-ones because they have lots of team meetings or they don't have enough time. Team meetings are good to promote interaction, but some team members are not as open or transparent. Identify the people who may require more frequent individual vs. team encounters. If the problem is time, a peer or an external coach can also give one-on-one coaching. Innovation and creativity is expected here to resolve the challenges, but one-on-one coaching should be part of the team routine.

Buddy programs[115] are also very effective, where you randomly pair people to get to know each other.

One-on-ones in Hybrid or Remote Work

Especially in times of uncertainty, a deep understanding of the employee's challenges and concerns is vital. Therefore, coaching sessions should be even more often.

In my interview with Jeff Schiefelbein, chief culture officer of the energy company 5,[116] he notes the company set a CARES team to connect with every employee each week. They do group coaching, supervisor coaching, and cross-coaching. Every month, they do a one-on-one deep dive, and annually they do a group coaching session that invites employees to develop the habit of self-reflection.

> *Watch the interview with Jeff Schiefelbein*
> *https://youtu.be/7yFxghwUQPs or download*
> *the We Culture app to watch more interviews.*

The company 1-800 Contacts also emphasizes coaching, now more than ever. Phil Bienert, chief marketing officer, said, "When working remotely, coaching has to be two or three times more frequent." Time management and work-life balance can be challenging, so employees need to be supported and coached to practice the right habits. Some employees may be alone and work longer hours; others may work at night to avoid family distractions.

When you're onboarding new employees, you need more consistent check-ins to ensure they know what's expected.

Leaders and middle managers should have conversations with their employees[117] more often than when working face-to-face, and not all of them should be focused on simply "did you meet your objectives or not." They should be comfortable also driving informal conversations about general well-being.

Especially when work is remote, feedback through email, or no feedback at all, can hurt employees' motivation tremendously. Concrete positive reinforcement has to be continuous.

The meetings don't necessarily have to take a long time. They can be 15-minute check-ins to replace random informal conversations you would have in the office.

Meaningful feedback to develop people's unique strengths helps employees on an ongoing basis and helps employees grow. If you are not comfortable doing it or don't know how to do it, look for coaching to do it better.

Most employees feel nervous during one-on-one online calls working from home in case the baby starts crying or the cat shows up in the background. In video calls, professionalism should be more about meeting objectives and being transparent about issues, not about mounting a perfect show. For instance, communicate that it is OK to go off-screen during a video call for a while to breastfeed. Likewise, communicate that it is also OK to see kids, pets, or family members in the background. Even though employees know it is OK, they may feel uncomfortable until they know it doesn't bother you and that you would do the same.

Hands-on 7.1

Organize frequent feedback sessions with your leader and teammates. Define a regular schedule and define a template (you can use the following GAP planning as a sample). Then, be ready to adjust the program in the future based on your experience or add extra sessions as needed. Document in your template the results of all the sessions, no matter how long or formal, to have a fair account of the improvement opportunities, and make sure you share the content with your co-worker.

Exercise

Think about your work, and what type of decisions you are able to make. Do you have a supervisor or co-worker who makes decision for you? How many of those decisions could now be solved by you? What do you need in order to make these changes?

Organize feedback sessions with your team.

1. Define a regular schedule.
2. Use this template as a reference.

GOAL-SETTING

- Purpose—what do you expect?

ASKING

- Alternatives?
- Ideas?
- What could go wrong?
- How to reduce challenges?
- Training needed?

PLANNING

- What
- Who
- When
- How

References

103. Rework with Google, "Hold effective 1:1 Meetings," https://rework. withgoogle.com/guides/managers-coach-managers-to-coach/steps/hold-effective-1-1-meetings/.

104. https://www.bls.gov/news.release/jltst.t04.htm.

105. Qualtrics, "The Other COVID-19 Crisis: Mental Health," *Qualtrics*, 2020, https://www.qualtrics.com/blog/confronting-mental-health/

106. BusinessSolver, "2021 State of Workplace Empathy," *BusinessSolver*, 2021, https://www.businessolver.com/resources/state-of-workplace-empathy.

107. Tara Van Bommel, "The Power of Empathy in Times of Crisis and Beyond," Catalyst, 2021, https://www.catalyst.org/reports/empathy-work-strategy-crisis.

108. Ben Wigert and Ryan Pendell, "The Ultimate Guide to Micromanagers: Signs, Causes, Solutions," *Gallup*, 2020, https://www.gallup.com/workplace/315530/ultimate-guide-micromanagers-signs-causes-solutions.aspx.

109. Ben Wigert and Jim Harter, "Re-Engineering Performance Management," *Gallup*, https://www.gallup.com/workplace/238064/re-engineering-performance-management.aspx.

110. Frederic Laloux, *Reinventing Organizations,* (Brussels, Belgium, Nelson Parker, 2014).

111. Frederic Laloux, *Reinventing Organizations* (Brussels, Belgium: Nelson Parker, 2014).

112. Karyn Twaronite, "The Surprising Power of Simply Asking Coworkers How They're Doing," *Harvard Business Review,* 2019, https://hbr-org.cdn.ampproject.org/c/s/hbr.org/amp/2019/02/the-surprising-power-of-simply-asking-coworkers-how-theyre-doing.

113. Richard Boyatziz, Melvin Smith, and Ellen Van Oosten, *Helping People Change* (Harvard Business Review Press, 2021).

114. Jennifer Jordan and Michael Sorell, "Why Reverse Mentoring Works and How to Do It Right," *Harvard Business Review,* 2019, https://hbr.org/2019/10/why-reverse-mentoring-works-and-how-to-do-it-right.

115. Luciana Paulise, "How to Lead a Remote Team with Vulnerability-Based Trust," *Forbes*, 2020, https://www.forbes.com/sites/lucianapaulise/2021/01/18/how-to-lead-a-remote-team-with-vulnerability-based-trust/?sh=78c97b7b988a.

116. Luciana Paulise, "Best Practices for Engagement and Productivity while Being Remote," *Medium*, 2020, https://medium.com/@luciana_19373/best-practices-for-engagement-and-productivity-while-being-remote-9a38151a478e.

117. Luciana Paulise, "How to Lead a Remote Team with Vulnerability-Based Trust," *Forbes*, 2021, https://www.forbes.com/sites/lucianapaulise/2021/01/18/how-to-lead-a-remote-team-with-vulnerability-based-trust/?sh=66c38d7a988a.

8

Skill #8: Promote Team Psychological Safety

Much of the work done at Google, and in many organizations, is done collaboratively by teams. The team is the molecular unit where real production happens, where innovative ideas are conceived and tested, and where employees experience most of their work. But it's also where interpersonal issues, ill-suited skill sets, and unclear group goals can hinder productivity and cause friction.

Following the success of Google's Project Oxygen research where the people analytics team studied what makes a great manager, Google researchers applied a similar method to discover the secrets of effective teams at Google. Code-named Project Aristotle—a tribute to Aristotle's quote, "The whole is greater than the sum of its parts" (as the Google researchers believed employees can do more working together than alone)—the goal was to answer the question: "What makes a team effective at Google?"

The researchers found that what really mattered was less about who was on the team and more about how the team worked together.[118]

What is Psychological Safety?

Amy Edmonson performed research on psychological safety in a hospital setting and published a paper in 1999 titled "Psychological Safety and Learning Behavior in Work Teams."[119] She defined psychological safety as "A shared belief held by members of the team, that a group is a safe place for taking risks."

In a team with high psychological safety, teammates feel safe taking risks around their team members, being disruptive, or inquiring instead of being seen as ignorant. They feel confident that no one will embarrass them for admitting a mistake, asking a question, or offering a new idea. This is vital to improving diversity because it can help make inclusion a reality.

When you feel psychologically safe, you take the risk to be yourself.

Amy Edmondson's results are aligned with later studies at Google, the Project Aristotle. When trying to find out what made teams effective, the researchers[120] found five things:

1. Psychological safety: Team members believe that a team is safe for risk-taking.

2. Dependability: Team members need to know they can depend on each other to complete quality work on time.

3. Structure and clarity: Teams members see clear goals and defined roles, that is, an individual's understanding of job expectations and the process for fulfilling these expectations.

4. Meaning: Team members feel their work is personally meaningful, and they feel a sense of purpose.

5. Impact: Team members believe their work is important. The results of one's work, thinking that it matters, is important for the team's cohesion.

Google discovered what mattered was less about who was on the team and more about how the team interacted, just as I mentioned with the team diamond structure: the strength of the interactions is what counts. A team of average performers could accomplish more than a team of stars when they interact effectively. The synergy effect, 2+2 = 5, takes place. And they also found out that the most important of all five characteristics was psychological safety.

Psychological safety has been studied in office workers, hospital nurses, and even astronauts, showing that psychological safety makes entire organizations perform better. The quality of one's relationship with one's teammates can impact performance, engagement, and innovativeness.

What is	What is Not
Allowing others to fail without repercussion	Never feeling completely relaxed because you don't want to make mistakes
Respecting diverging opinions and welcoming conflict as a way to find together the best option	Being critical of each other
Feeling free to question	People trying to show authority by speaking louder. Nobody listens, everyone wants to be the leader
Trusting that people are not trying to undermine you	Feeling like I have to prove myself

The Impact of Psychological Safety (or the Lack of It)

Imagine you have an idea, a new way of doing something. How would you feel about proposing the idea? Would you be nervous? Excited? Do you think anyone would listen to you? Would you dare to define the status quo? How do you think your supervisor would react?

People tend to avoid doing things that can negatively influence how others see them. While this is a form of self-protection, it can negatively impact how people interact in a team setting. People who do not feel psychologically safe avoid proposing ideas or speaking up if they think there might be a problem in the process.

Psychological safety encourages people to be themselves and speak freely, promoting the development of ideas and creativity and, consequently, increasing employee commitment. All these behaviors help the company promote continuous improvement in the long term, impacting the bottom line, performance, engagement, and innovation. Innovation increases while turn-over rates and absences decrease significantly.

Unfortunately, psychological safety is not yet common in the workplace, and it is even more challenging to build in a remote environment.

> *Diversity can be created through deliberate hiring practices; inclusion does not automatically follow. Having a diverse workforce most certainly does not guarantee that everyone in your organization feels a sense of belonging. In particular, when no one at the top of the organization looks like you, it makes it harder to feel you belong. It is difficult to feel a sense of belonging when one feels psychologically unsafe.*
>
> –Amy Edmondson[121]

A recent example of the impact of a psychologically safe environment managed by the numbers was the Wells Fargo case. The company's strategy was to push its employees on cross-selling. Around 2015, employees had the sense that they would be fired if they didn't achieve the set targets. Managers were very tough and present. People did not feel it was safe to push back or say that the strategy wasn't working or the targets could not be met.

Wells Fargo ended up encouraging behaviors in its employees that were not supposed to be encouraged, pushing beyond the customer's possibilities. In 2015, the city of Los Angeles sued Wells Fargo for unethical customer conduct,[122] accusing the bank of secretly opening unauthorized accounts that then accrued bogus fees. The company was fined $185 million for opening the unauthorized accounts between 2011 and 2015 as a way of bumping up sales figures.

Amy Edmondson analyzed psychological safety in a hospital. For example, a nurse in one team explained, "Mistakes are serious, because of the toxicity of the drugs [we use] so you're never afraid to tell the nurse manager"; in contrast, a nurse on another team in the same hospital reported, "You get put on trial! People get blamed for mistakes . . . you don't want to have made one." In the first team, members believed that speaking up was natural and necessary, and in the other, speaking up was viewed as a threat.

Organizations can define targets and strive for error-free processes, but employees should be part of the discussion. They are with the customers in the day-to-day; they build the products or deliver the services. They need to know that they can say when something is wrong. When team members make a mistake or realize a process is misleading, unsafe, or not helping the customer, they know it. So, it's up to the organizations to create a safe space for employees to speak up and leaders to be active listeners to do something about it.

Blaming others and always agreeing with the boss or the "hippo" (the high potential in the group, whether it be a leader or the highest authority) are also signs of lack of psychological safety.

Me Culture Behavior: Not Speaking Up When Managers or High Performers (aka Hippos) Are Present

It is not surprising to see how employees react when a manager is speaking or simply present in a meeting; they usually nod and avoid challenging the manager. It is the opposite in companies where psychological safety is key. Everybody feels safe to challenge anyone in the company; you never know who can think of the best ideas.

Pixar is a good example. Ed Catmull, co-founder and leader, was very deliberate in creating a psychologically safe environment where candor is expected—possible critical feedback—especially during the company's brain trust meetings. These were frequent feedback sessions essential for critiquing movies in the making, where everyone had to be able to talk and offer candid feedback. Candid feedback means that if you find a mistake or something that can be improved, you will disclose it, and everyone has to take it positively. Catmull says in his book, *Creativity Inc,*[123] that without the freedom to fail, people "will seek instead to repeat something safe that's been good enough in the past. Their work will be derivative, not innovative." We need to understand failure as a natural part of learning and exploration.

Steve Jobs was one of the three founding fathers of Pixar Animation Studios. Imagine if your company is based on the idea that some are much better than others; how would you feel if you had Steve Jobs in the company? You would probably be led to believe he is the most important and know-it-all person in the room. Still, Pixar's rule was candor during meetings. So if Steve Jobs was preventing other people from being candid, then he wouldn't be in the meeting, at least not the entire session. Yes, they would kick Jobs out of some meetings. What happens is that no matter how much you ask employees to be candid, they may still avoid speaking or being controversial when a manager or a figure such as him who is the hippo or hipper potential is onsite.

The lack of psychological safety fosters that ideas are not challenged when they are based on leaders' beliefs, and this is the worst-case scenario for idea generation and problem-solving you can have. What you want is for all employees to feel free to speak up, criticize, and propose solutions. Then, after everyone has participated, you can let the managers participate. As a consultant, I have worked with this principle multiple times. When feeling confident and in an environment of trust, employees alone would discuss serious things openly that would later on drive to amazing improvements. And besides, employees would also feel less burdened and more accomplished being able to speak their minds and contribute.

We Culture Behavior: Create a Safe Space

Creating a safe space for your team is the most basic human need. It is the second most important in Maslow's hierarchy of needs. Crisis especially makes it even more challenging. A lack of communication can quickly erode trust and psychological safety. Organizations have to set up the right 7Rs across all the steps of the employee experience (see the end of this chapter) in order to drive psychological safety.

Below are some recommendations to increase the psychological safety of virtual teams:

Meetings. Create an environment that makes people comfortable making the right decisions, asking questions, and being themselves in every meeting. Set up sessions so it is easier to give each other candid feedback by including a round-robin during the meeting, or at least to including everyone's comments at the end. Some may not feel comfortable speaking up in public; you may want to identify these people and work with them after meetings or coach them to be more open. Ask them how they feel they could feel more comfortable.

One-on-one meetings. Schedule frequent check-ins with employees. Try to organize with every employee at least every week to discuss performance and how he or she is feeling. See "Chapter 7, Build Empathetic 1:1 Relationships."

Happy hours. Organize virtual happy hours or events to increase casual interaction among team members. In a Zoom interview with the company 97th Floor, employees connect for lunch every day. These meetings have no agenda; everybody can dial in. It's a chance for employees and management to connect. In another interview with 1-800 Contacts, CEO John Graham mentioned the company hosts a virtual lunch on Tuesdays and virtual coffees on Thursdays.

Watch the interview with
Annalee Jarrett with 97th Floor
https://youtu.be/ypVx-W7V2e4

Open channels. Team members may feel more comfortable asking questions or proposing solutions in smaller teams. Use channels to discuss topics. For example, some companies use Slack to focus on specific issues or share resources with like-minded people.

Equality groups. Affinity groups, as discussed in the previous chapter, are another way to provide a safe space for people to connect around various aspects of their identity, culture, or interests, such as women, Hispanic, or African American groups. They can have constructive and psychologically safe conversations in smaller teams.

Meditation and mindfulness. You can hire an expert to offer virtual sessions, or you can try starting a team meeting with a guided meditation. They are used to boost emotional intelligence, reduce stress and anxiety, and increase resilience.

Debate ideas productively. Psychological safety might be less efficient in the short run, but it's more productive over time. Challenges don't need to be avoided to make people feel safe. Instead, they will be well received as long as they are not threating. Google's environment, for instance, is highly demanding and innovative. It is precisely in this type of environment where the ability of people to feel comfortable taking risks, deciding fast, and participating is most needed. Biology shows that a challenging environment without being threatening helps increase oxytocin in the body. This hormone is a neurotransmitter that is involved in behaviors related to trust, altruism, generosity, bonding, caring behaviors, empathy, and compassion, and in the regulation of fear, eliminating paralysis responses.

Measure psychological safety. Use anonymous surveys and pulses to measure how people feel about taking risks and bringing their whole selves to work. The following survey designed by Amy Edmondson can help.

Survey
The answers marked with (R) imply
a risk for the employees.

- When someone makes a mistake in this team, it is often held against him or her (R).
- In this team, it is easy to discuss difficult issues and problems.
- In this team, people are sometimes rejected for being different (R).
- It is entirely safe to take a risk on this team.
- It is difficult to ask other members of this team for help (R).
- Members of this team value and respect each others' contributions.

Organize empathy workshops. For example, the Arnold and Havas Media company holds empathy workshops.[124] Team members draw from a card deck with self-reflective questions such as "What makes me unique?" Before answering, the team member must articulate their internal reaction ("my heart is beating faster," "my mind just went blank") and label that emotional response ("embarrassment, anxiety"). Making these personal disclosures builds empathy in the other participants before they hear the answer to the question.

After the speaker concludes, the listeners must also share *their* emotional reactions and answers to the card questions. As one participant said to a speaker during the exercise, "I saw you getting anxious, and then *I* got anxious because I was putting myself in your shoes imagining how I would answer that question."

Set the example. In Edmondson's hospital studies, the teams with the highest levels of psychological safety were also the ones with leaders who were most likely to model listening and social sensitivity. She says, "The best tactic for establishing psychological safety is the demonstration by a team leader." Leaders have to show that they can make mistakes." If the leader is supportive and has nondefensive responses to questions and challenges, team members understand they can respond the same way too. But if team leaders punish asking questions or being delayed, team members will be reluctant to share losses, mistakes, or questions.

Invite engagement. Leaders can invite engagement by saying things like, "We need to hear from you. What ideas do you have? Let's test them." This attitude then will be copied by other team members even when leaders are not present.

Respond productively. It is one thing to drive psychological safety when you want to ask for feedback and ideas, as you are in a positive environment. But how do you keep people safe even when they make mistakes? What happens when one is delayed on a project? A productive response would be thanking the person for being open about it and asking how you can help get it right. Employees should feel safe to say when something is wrong and feel relieved that they will get the help they need no matter what. In case the delays and mistakes are too often, the employee should be coached about their work processes, habits, or time management skills and identify the root cause of the frequent delays. The "how the work is getting done" should be discussed in these particular cases.

Toyota, for example, designed the Andon cord in the production line. Employees are expected to pull that cord when there is something that is not OK. It is part of the production process, so nobody feels ashamed of it; it has been normalized. The Andon cord also plays music that alerts others that a team member is having issues, and when the problem has not been solved timely, the line stops.

Estimate the internal cost of psychological safety. Employees who work in companies that are not psychologically safe tend to have a hard time getting used to stopping a process, being candid, or urging others to be honest with them. Estimating how much money can be left at the table when an idea is not proposed or a mistake is not highlighted can help put shame in perspective. For example, employees know that a new car is made at a Toyota plant every minute. If the line stops for one minute, everybody knows how much money is lost. But they also know that not stopping to fix the problem only makes it more expensive. A recall, an accident, or a customer complaint can be many times more expensive.

Belonging queues enhance communication. For psychological security to emerge among a group, teammates do not necessarily have to be friends. However, they must be socially sensitive and ensure that everyone feels heard. Charles Duhigg suggests in his book *Smarter Better Faster*[125] that "belonging" queues from the leader and team members help people feel safe in the team. Some examples are:

- Active listening
- Eye contact
- Frequent interactions
- Vulnerability shown by raising concerns, admitting mistakes, asking for help or feeling sorry
- Continuous feedback
- Team cohesion during hard moments
- Hugs
- Frequent feedback
- Speaking the same amount of time

Offer autonomy based on performance. Timothy Clark,[16] in his book *The 4 Stages of Psychological Safety*, recommends that leaders offer autonomy as a way to drive psychological safety. Workers should have the chance to contribute to their teams in self-directed ways; at the same time, the workers must accept the responsibility to perform. While some leaders may be hesitant to give power to others, the process behind it is feeling diminished or not needed. In this case, leaders need to be coached to gain security and psychological safety themselves.

Managing Virtual Meetings

You can foster the same communication and transparency while working remotely.

Virtual meetings should be seen as moments to analyze results, share experiences, and look for solutions as a team; they should not be used to punish unexpected results. If employees feel threatened to show a negative result, they may be tempted to hide it, sugar coat it, or manipulate results instead of asking for help. Everyone should be invited to speak. The following are some tips to consider:

- **Communicate.** Communicate with your staff regularly and consistently, at the same time, to maintain a routine and ensure a safe space. Invite all members to turn on the video and maintain eye contact as much as possible. To increase engagement, reduce meetings to 30 minutes or less, invite every member to speak up, and use all the tools possible to communicate. Given that more than 60% of the communication is nonverbal, use video, hands up, emoticons, surveys, polls, breakout rooms, or digital whiteboards.

- **Breakout rooms.** Creating smaller virtual breakout rooms during large meetings allows small groups of three to five people to dig deeper into certain subjects. They provide a psychologically safe space to test ideas and build relationships.

- **Hand-raise, yes/no and chat.** Allowing everyone to contribute at the same time through these tools is an option, but it needs to be managed carefully. Not everyone is able to respond quickly or may not feel comfortable to be exposed, so results may look like a false positive. Take your time to allow everyone to provide their opinion.

- **Polls.** Edmondson[127] strongly recommends using polls to "make it easy to express an opinion without fear of being singled out and prompt thoughtful probing to dig into diverse views."

If the group norms and values in Skill #1 are not defined or ingrained in the company's culture, unwritten rules will take over, for good or bad. Groups develop norms about appropriate behavior. Some norms correlate with highly effective teams and others don't. Psychological safety is one of the positive ones. But, if most team members don't feel safe speaking up, this will become the norm, whether written or unwritten. The bottom line: psychological safety is not optional; if it is not intentional, the lack of it will erode the team's effectiveness.

Hands-on 8.1

Team exercise: Share the following chart with your team and ask them to fill it in anonymously, thinking about how the specific team works. Analyze the answers and discuss how they feel about it, and find out together a way to improve psychological safety within the team.

Team Exercise

Share these questions with your team and ask them to fill it in anonymously about how they feel the team works.

STATEMENT	True	False
When you make a mistake it is often held against you.		
It is easy to discuss difficult issues and problems.		
People are sometimes rejected for being different.		
It is safe to take a risk, provide ideas, or try.		
It is difficult to ask other members of this team for help.		
Members of this team value and respect each others' contributions.		

Hands-on 8.2

Individual exercise: Think about moments when you did not feel safe at work or when you made others feel unsafe. Come up with three ideas on how to improve psychological safety at work. What can you do to improve? What can your team do to improve?

. . . .

Individual Exercise

Think about moments when you did not feel safe at work, or when you made others feel unsafe. Come up with three ideas on how to improve psychological safety at work.

_____

_____

_____

_____

_____

. . . .

References

118. "Guide: Understand Team Effectiveness," Rework with Google, https://rework.withgoogle.com/print/guides/5721312655835136/.

119. A. Edmondson, "Psychological Safety and Learning Behavior in Work Teams," *Administrative Science Quarterly* 44, no. 4, December 1999, 350–383.

120. Guide: Understand Team Effectiveness," Rework with Google, https://rework.withgoogle.com/print/guides/5721312655835136/.

121. Amy Edmondson, "The Role of Psychological Safety in Diversity and Inclusion," *Psychology Today*, 2020, https://www-psychologytoday-com.cdn.ampproject.org/c/s/www.psychologytoday.com/us/blog/the-fearless-organization/202006/the-role-psychological-safety-in-diversity-and-inclusion?amp.

122. Maggie McGrath, "Wells Fargo Fined $185 Million For Opening Accounts Without Customers' Knowledge," *Forbes*, 2016, https://www.forbes.com/sites/maggiemcgrath/2016/09/08/wells-fargo-fined-185-million-for-opening-accounts-without-customers-knowledge/?sh=77cb0dde51fc.

123. Ed Catmull with Amy Wallace, *Creativity Inc.* (New York: Random House, 2014).

124. Constance Noonan Hadley, "Employees Are Lonelier Than Ever. Here's How Employers Can Help," *Harvard Business Review,* 2021, https://hbr.org/2021/06/employees-are-lonelier-than-ever-heres-how-employers-can-help.

125. Charles Duhigg, *Smarter Faster Better: The Transformative Power of Real Productivity* (New York: Penguin Random House, 2016).

126. Timothy R. Clark, *The 4 Stages of Psychological Safety* (Oakland, CA: Berrett-Koehler, 2020).

127. Amy C. Edmondson and Gene Daley, "How to Foster Psychological Safety in Virtual Meetings," *Harvard Business Review,* 2020, https://hbr.org/2020/08/how-to-foster-psychological-safety-in-virtual-meetings.

9

Skill #9: Embrace Diversity, Equity, and Inclusion

We had a diverse organization, but we didn't have inclusivity right. Focusing on both is just the right thing to do, and we've all been struck by how much better it is when a diverse squad becomes truly inclusive. They know how to work together as a group, every voice comes to the table, and it's extraordinary how much better the outcomes are and how much better the workplace feels. Because squads are limited to no more than 10 people and have a clear mission and purpose, everyone has to have a voice. There isn't a place for anyone to just cruise along.

–Spark New Zealand[128]

B uilding a diverse team is about having people with various skills, experiences, and backgrounds that enrich each other to achieve more together than individually. Different people with different profiles, points of view, and knowledge are the fuel that can take a company above and beyond.

The skill of embracing diversity is developed not just by having a diverse workforce at the company level, but also when diversity is welcomed and fostered at the team and individual levels, making the individuals feel safe in their inner circle.

We tend to form teams with people similar to ourselves. For starters, most companies are still organized by departments, where you have teams of accountants, executives, engineers, sales associates, and so forth.

But there is always a situation where you may feel like a toad from another well. For example, you may be a millennial in a company led by baby boomers, a woman in a shareholder's meeting full of white men, a black nurse on a team of white doctors, or an LGBTQ+ in a heterosexual, all-male team. All of these are situations where people feel different due to race, gender, experience, age, or an exceptional circumstance they need to deal with.

We are all different in certain aspects and still similar in others. The problem is not to be different but to be treated differently or have access to fewer opportunities.

Usually, achieving diversity, equity, and inclusion (DEI) is not a one-time project implementation. You may achieve pay equality today, but unconscious bias will bring inequality back again next year.

As per a BCG[129] research, while the United States alone spends an estimated $8 billion a year on DEI best practices, only half of the diverse employees feel there are mechanisms in place to ensure decisions are free from bias.

It is a moving target, but you can keep it under control with periodic employee surveys.

Diversity also varies substantially across business units within a company, so to achieve the real benefits that diversity can bring, leaders must check the gender balance in specific business units too, not only at the global level. Diversity is also important in procurement, as opening the game to smaller and diverse businesses can bring new and different services, significant price reductions, and more customized proposals.

To improve diversity and equality, companies need to update their policies, strategies, cultures, and values to be able to attract and retain minorities.

Multidisciplinary Teams

The health industry has been benefiting from multidisciplinary teams. Usually, teams have workers from different disciplines such as nurses, surgeons, psychiatrists, or social workers, each providing specific services to patients without consulting other teams. They coordinate their services and work together toward a specific set of goals.

The research below shows that building multidisciplinary teams has a tremendous impact on the bottom-line results in other industries too. That's how agile companies are organized.

When staff members are widely cross-trained (creating redundancy), then if one operation is overloaded or some positions are missing, others can step in to help, making the team more agile to respond to the change.

The best teams are the most diverse and include people from different studies, sexes, races, and social statuses. In the teams, every member must be respected. Their contributions are heard to increase synergy, that is, the ability of the team to achieve better results than the sum of individual efforts. When a team does not work well, not only is it not synergistic, but the result is also less than the sum of the individual efforts.

The role of the agile leader is to create an environment where all people can do their best.

For example, an entrepreneur can be a generator of work for himself, but if he wants to make a company, he must form a team. First, leaders must know themselves well (strengths and weaknesses), know their role, and then find a team that complements it.

Multidisciplinary teams tend to be self-sufficient, as they possess many different talents that can conquer most problems. When working together, they tend to learn from their peers, who all have a unique set of skills to offer and complement each other.

Although communication can be challenging in a multidisciplinary team, it will work best when team members practice empathy and actively listen when sharing points of view among peers.

What is Diversity, Equity, and Inclusion?

Learning that diversity, equity, and inclusion are not the same is essential to identifying where the company lacks more action.

Diversity means you have the presence of differences of identities (for example, gender and people of color) throughout your organization. *Equity* is an approach that ensures everyone has access to the same opportunities. The company must make sure everyone has the opportunity to grow, contribute, and develop,[130] regardless of their identity. Finally, *inclusion* is about people with different identities feeling valued, leveraged, and welcomed in the company.

Increasing diversity or increasing the representation of certain demographic groups is an objective all companies should strive for. Still, the company must also focus on enabling these groups to contribute (inclusion) and advance in their careers (equity).

Equity looks forward to achieving unbiased treatment. Some groups have more advantages than others, so equity efforts try to correct such differences.

Unfortunately, social identity categories such as gender, race, ethnicity, sexuality, age, ability, and status consist of dominant and nondominant groups. People in dominant groups tend to have more economic power and opportunities than those in nondominant ones. Nondominant groups may experience discrimination and limited opportunities. For example, men in most societies are the dominant gender group, and all other gender classifications are nondominant. They tend to have more opportunities to grow at work, especially at higher levels. Statistics show that fewer women are CEOs than men. According to *Catalyst*, women currently hold only 6% of CEO positions at S&P 500 companies.[131]

Diverse and inclusive teams should promote the same processes for everyone regardless of gender. It is also essential to understand that diversity is not about people based on race or orientation, but about people with different needs. Certain demographic factors such as age, being an immigrant, being a caretaker for someone with cancer, or mental health challenges are examples of situations where employees may experience work differently and, therefore, need more flexibility. Leaders should understand that anyone can be in any of these situations, so addressing diversity is critical to ensuring they can help their teams work best no matter their circumstances.

Me Culture Behaviors: Fighting for Personal Advantage

Companies can say they are diverse or inclusive, but if employees don't feel that way, both the company and the employees are being impacted. As a result, companies miss out on hiring a portion of talent, and the communities don't have the same career development opportunities.

People who don't feel welcome in the company contribute but in a more limited way. They may not participate fully in their jobs or may resent those who receive attention. They may be sick or feel depressed more often. They may not contribute in meetings, or if they do, their contributions may not be heard. There is also the concern about the nondominant group that may not be on board with being inclusive, and their resistance could slow down progress toward inclusion.

Unfortunately, the fight for benefits and personal advantage contributes to a system that favors being the same as opposed to being different. Being different is risky. If you're benefiting from the current system, you're likely to resist changing it, so majorities are not going to help minorities.

For instance, MIT research[132] suggests that in the United States, racism is systemic. White employees have benefited from a personal characteristic over which they have no control, so systemic racism is challenging to acknowledge. And the worst part is that being systemic is part of every aspect of organizational life, from hiring to developing to evaluating.

The MIT research explains that "individuals can't simply create an inclusive working environment for themselves. We're seeing that colleagues, leaders, and talent processes all play important roles in promoting inclusion."

So the impact of not having a team that values diversity is that the team members are not unleashing their highest potential, which makes them willing to leave the organization. As a result, the team doesn't take advantage of the synergy effect, and the company misses the opportunity to explore new perspectives.

How is This Lack of Inclusion Affecting the Business?

A McKinsey study[133] showed that companies in the top quartile for racial/ethnic diversity were 35% more likely to have financial returns above their national industry median. The study reveals that the most important drivers identified were advantages in recruiting the best talent, stronger customer orientation, increased employee satisfaction, and improved decision-making.

In addition, according to the previously mentioned BCG research,[134] diversity has proved to boost innovation and strengthen resilience. For example, companies that reported above-average diversity on their management teams also reported innovation revenue that was 19 percentage points higher than companies with below-average leadership diversity — 45% of total revenue versus just 26%. On the other hand, the impact of not being diverse translates to a smaller talent pool, more hiring costs, and less collaboration, among others.

Key Insight
Diversity has proved to boost innovation and also strengthen resilience.

These are some of the benefits:

- **Increased psychological safety for everyone.** You can have a diverse team of talent, but that doesn't mean you feel valued or psychologically safe to speak your mind. Psychological safety was found to be one of the major factors for a team to be successful. It means that all team members feel safe to be themselves; therefore, they can unleash all their talent. Everyone can feel threatened. Stress is significantly reduced when the environment promotes respect and support for others. (See Skill #8, Promote Team Psychological Safety.)

- **More points of view bring more innovation.** The most innovative companies value diversity. Especially in a company where 90% of the employees are engineers, such as in technology companies, you can bring different points of view by hiring people from different origins, cultures, and languages and boost innovation. Diversity fosters creativity. Accepting new perspectives brings more solutions and different problem-solving approaches. Decisions are made faster.

> *Innovation cannot advance in a positive direction unless it's grounded in genuine and continued efforts to lift all humanity.*
>
> –Marc Benioff[135]

- **Increased talent pool.** Diversity in leadership helps companies access more sources of talent. The skill gap is intensifying, so now the market is employee-driven. Diversity management is key to addressing talent shortages and giving companies an advantage when competing for the best talent. On the one hand, minorities will look for companies that respect them and give them the same opportunities. On the other hand, once your workforce is trained to reduce their biases, they will be able to find talent in individuals whom they would have ignored in the past. Besides, the groups targeted by diversity efforts are often good sources of desirable talent. A recent study[136] found that on average, LGBT+ recruits were more highly skilled and more likely to have advanced degrees.

- **Strengthened customer orientation.** Diverse companies align better with a diverse customer-base because all ideas are heard within the teams. Women and other minorities are key consumer decision-makers and bring new perspectives to the table to offer better products and services.

- **Improved decision-making.** Meetings are more effective as the teams work better together. Communication is more transparent and direct, providing more information to make decisions. Accepting new perspectives brings more solutions and different problem-solving approaches. Decisions are made faster.

- **Enhanced company image.** As social responsibility becomes more important, customers and employees will weigh their decisions based on the company's image. Salesforce, for example, is on the list of *Fortune's* best companies to work for, and it is on *People* magazine's list of "Companies that Care."

- **Increased employee engagement.** When all employees feel psychologically safe to speak up, conflicts between and across groups are reduced, and stress is reduced because the "winning at all cost" mindset is replaced by a collaboration and "fail fast" mindset. Employees don't feel threatened; on the contrary, they feel valued and therefore are more willing to engage. Salesforce, led by CEO Marc Benioff, also fosters employee engagement through a core value — equality. One of its ways to promote equality is to encourage employees to join their equality groups: "affinity groups that provide a safe space for people to connect around various aspects of their identity" where they can have constructive and psychologically safe conversations.

Me Culture: Perpetrating Barriers to DEI

There are many barriers within companies that prevent people from feeling equal. It's time to review the entire employee experience of a minority and identify process weaknesses. Statements are what we say we do, but again, culture is what we really do. Companies need to build DEI into their culture. The real benefits of DEI start when new behaviors are learned and the company culture changes to prioritize respect over power.

Key Insight

The real benefits of diversity, equity, and inclusion start when new behaviors are learned and the company culture changes to prioritize respect over power.

Biases

Unconscious bias is the tendency of people to trust and favor those who look like them. Unfortunately, fighting unconscious or implicit bias before it occurs is very difficult because it operates beneath our awareness. Perhaps that's why, on their own, diversity training programs have produced mixed results and sometimes have even been found to make things worse.

Even though I am a woman, I may unknowingly treat women differently than men. This can be due to unconscious bias. People tend to be blind to their biases, favor people who look or behave like them, or favor the majority's preference. Many times even women prefer men due to this.

We all do that no matter our ethnicity. Training and coaching the workforce is vital to reduce these biases.

Racial bias

If you have a lake in front of your house and one fish is floating belly-up dead, it makes sense to analyze the fish. But if you come out to that same lake and half the fish are floating belly-up dead, what should you do? This time you've got to analyze the lake. Now...

picture five lakes around your house, and in each and every lake half the fish are floating belly-up dead! What is it time to do? We say it's time to analyze the groundwater. How did the water in all these lakes end up with the same contamination? On the surface the lakes don't appear to be connected, but it's possible— even likely—that they are. We live in a racially structured society, and that is what causes racial inequity. In other words, we have a "groundwater" problem, and we need "groundwater" solutions.

–The Groundwater approach
by the Racial Equity Institute

A We Culture thrives when the experience of white people is as important as the experience of those who identify as black, indigenous, and people of color (BIPOC). Unfortunately, racial inequity seems to impact every system in the United States—in healthcare, education, law enforcement, child welfare, and finance, to name a few. For example, the Racial Equity Institute[137] reports in a study from 2018 that "African Americans and Latinos continue to be routinely denied conventional mortgage loans at rates far higher than their white counterparts."

In a 2015 study of education and discipline, Stanford psychologists Jennifer Eberhardt and Jason Okonofua presented teachers with written vignettes of student misbehavior. The vignettes were identical except that half had "black-sounding" names and half had "white-sounding" names. Teachers of all races said that (fictitious) students with black-sounding names were more disruptive, more likely to be repeat offenders, and more appropriately labeled as "troublemakers."

The same happens at work.

Black workers are underrepresented in the highest-growth geographies and the highest-paying industries. Meanwhile, they are overrepresented in low-growth geographies and in front-line jobs, which tend to pay less.

Forty-three percent of black private-sector workers earn less than $30,000 per year, compared with 29% of the rest of the private sector.[138]

Black women and Latinas also face more barriers to advancement than most other employees. For every 100 men promoted to manager, only 58 black women and 71 Latinas are promoted. Latina mothers are 1.6 times more likely than white mothers to be responsible for all childcare and housework, and black mothers are twice as likely.

After the pandemic there were also numerous anti-black and anti-Asian events that resulted in the loss of life, injuries, and psychological harm, such as the death of George Floyd or the shooting in north Georgia in March 2021.

In the United States, organizations have started celebrating Juneteenth on June 19 to commemorate African American Independence Day, especially after George Floyd death. Many organizations also celebrate National Hispanic Heritage Month from September 15 to October 15, as Hispanic and Latino Americans amount to an estimated 17.8% of the total U.S. population, making up the largest ethnic minority. Still, these celebrations are not enough; racial bias has to be understood and looked in the eye every day to be able to reduce it, one interaction at a time.

Microaggressions

Some instances that reveal people's intrinsic biases leave their listeners feeling uncomfortable or insulted. These *microaggressions* are small messages that usually members of nondominant groups experience from individuals who are often unaware that they have said or done something offensive. For example, statements like "Your English is great" to an immigrant or questions like "Is that your real hair?" can make people feel uncomfortable in subtle ways.

These include verbal and nonverbal cues such as words, facial expressions, and vocal tones. For example, a person rolls their eyes when another person speaks with an accent. Another example is racially profiling one member of an under-represented group with another person from the same group, assuming that all Hispanics are from Mexico. One last example is, when you meet two coworkers, one male and one female, assuming that the male co-worker is in a senior position rather than his female colleague and therefore asking questions or addressing mostly to him This also happens a lot with couples. When I am with my husband, in many cases people we meet tend to ask him where he works and avoid asking me. Maybe they are assuming I don't work? I don't want to assume either.

As these messages are subtle, the individuals may not realize their negative impact, and they may even diminish it, saying the listener is overreacting. However, if these aggressions happen too often, they compound over time and can have adverse effects on employee experience, health, and mental well-being.

Onlyness

Being the "only" on a team, a virtual meeting, or a room given your gender identity, sexual orientation, or race (for example, being the only woman in a meeting full of men) drives anxiety, resulting in more pressure to perform to reinforce the difference and a sense of being on the spot.

A McKinsey report[139] says that women are more likely to experience discrimination in the workplace than men. But the study shows the odds are higher still when women find themselves alone in a group of men. They are far more likely than others to have their judgment questioned than women working in a more balanced environment (49% versus 32%), to be mistaken for someone more junior (35% versus 15%), and to be subjected to unprofessional and demeaning remarks (24% versus 14%).

Onlyness impacts organizational performance and the individual performance of those who feel the only, while eroding psychological safety. It is hard to feel safe when you are on the spot. Some of these individuals can still succeed, but others may decide to leave the job or stay quiet.

To increase diversity, companies may distribute one "only" in every group. The McKinsey Research suggests that "it is preferable to assemble teams comprising several women rather than try to place one woman on every team" in order to support each other. Companies need to look for ways to prevent onlyness.

Training leaders and "onlys" to find common ground among team members instead of differences is another way to avoid a negative connotation. An only can really benefit a team by offering new perspectives and ideas (as long as new ideas are welcomed) while proving to have the same technical expertise. Leaders should advocate for them and become sponsors to reduce the team bias whenever possible.

Different Generations

Being a different age is also a source of discrimination. Groups of people of similar ages are called generations. Baby boomers are those who were born in 1945-1963, Generation Xers were born between 1963-1980, millennials were born between 1981-1995, and Generation Zs were born after 1995. Every generation has different values, methods of communicating, and ways of feeling motivated.

Gen Xers complain that millennials are not committed,[140] millennials complain that Gen Xers are not passionate and work too hard, baby boomers want to call Gen Z on the phone, but Gen Zs want to reply by sending a TikTok video.

The problem is that different generations have other working habits, which makes it difficult to sustain a culture across the organization regarding customer service, safety, and continuous improvement. This disconnection or disengagement impacts the organization, resulting in high absenteeism rates and low retention, as per the Gallup report.[141]

A survey by Pew Research Center[142] shows that 10,000 baby boomers retire each day. This shift in demographics is transforming the workforce. As more boomers enter retirement, Gen Z will be replacing them, bringing with them an entirely different worldview and perspective on their careers and how to succeed in the workplace. Whatsmore, generational differences also highlight differences in how long people stay in their jobs.[143] With millennials and Gen Z leading the trend of "job hopping," not surprisingly, in early 2021 during the pandemic, it started the Great Resignation where companies started to see higher rates of people migrating to other organizations.

Baby Boomers

Baby boomers, born between 1946 and 1964, tend to stay in a job for an average of eight years and three months. Various events impacted this generation while they were growing up: The World Wars ending, leading to a sense of optimism and prosperity; landing on the moon; the civil rights movement; and the idea of living the American dream.

Because of these events, when baby boomers had a job that paid well and allowed them to live comfortably, they didn't feel the need to leave. A top-down direction is very much preferred by baby boomers, while it is not as welcome for younger generations, who want to be part of decision-making. Baby boomers don't want to be perceived as old or aging. Finally, they are considered the "me generation," and their values are justice, integrity, family, and duty. Facebook is their most common platform, and they are motivated by public recognition.

Generation X

Born between 1965 and 1980, Gen Xers will spend a little more than five years in a job.

Some of the cultural events that have impacted Gen X's attitudes to work are: receiving the highest level of education in the United States of any previous generation; the fall of the Berlin wall, symbolizing freedom and victory; the start of dual-income families; and the energy crisis.

Gen Xers are self-sufficient, resourceful, and individualistic, since they became accustomed to caring for themselves as moms started to work. They value freedom, but mostly they are responsible. They respect the rules and their leaders, and prefer to be given the autonomy to work under minimal supervision. They are ready to work as many hours as needed to get the job done.

They tend to be problem-solvers, analytic, and strategic thinkers. Because they lived through difficult economic times in the early 1980s, they are less committed to employers than baby boomers, but they are more loyal than the younger generations. Gen Xers were the first generation to grow up with computers and to learn and adapt to new technologies. Family, autonomy, and stability are their values.

Millennials

Millennials were born from 1981 to 1995. Millennial workers stay in a job for an average of two years and nine months. With about 72 million millennials in the United States, this generation is the biggest of the working population. The events that occurred while they were growing up that impacted their behavior were: The Great Recession from December 2007 to June 2009, impacting wages and cost of living for years; advancements in technology; the increase of terrorism; and the high divorce rates.

They like to stay connected 24/7 and spend approximately 24 hours a week searching online. Within this generation, 80% sleep with their phones next to their beds, and 27% are self-employed. They want to buy products that create a great buying experience for them. They don't just a phone, they want to go to a massive Apple store or check it out from home through augmented reality. They want to explore the product, watch reviews, and customize it. They are entrepreneurs. They want to say what they think. Their values are happiness, passion, diversity, and sharing, which is why they are called the "we generation."

. . . .

*With Generation Y coming into the business,
hierarchies have to disappear. Generation Y expects
to work in communities of mutual interest and
passion—not structured hierarchies. Consequently,
people management strategies will have to change
so that they look more like Facebook and less
like the pyramid structures we are used to.*

–Vineet Nayar, Vice Chairman and
CEO, HCL Technologies, India[144]

. . . .

They say they don't want to stay for a long time at the same company; they want to move frequently to different positions. They want to be heard. They like freedom, so millennials are rethinking their work-life balance, especially after the pandemic, and want to take advantage of the remote work policies.

Millennials are stepping into leadership positions while Gen Zs are entering the workforce.

Generation Z

Gen Zs were born between 1996 and 2012. The average time spent in a job is bound to be less than the generations before them. Gen Z's average length in a job (so far) is two years and three months—only six months less than millennials, but it is important to consider that many haven't even had the chance to be in a job for more than three years yet, as some are just entering the workforce.

Most Gen Zs were born after 9/11. They grew up in a hyper-connected world where the cell phone is more important than the computer, and where the rewards seem to be instant, explaining why they don't stay in jobs very long. However, Gen Z is much more cautious with risk-taking than millennials.

They don't simply consume content online like millennials; they love to create content. Almost 25% post videos almost once a week, 65% post all types of content, and nine out of ten watch YouTube daily. They are multitaskers.

Some of them are called "smart creatives," and many of them are employed in the tech industry. They are creative and self-directed, collaborate freely, and like to express themselves. As the book *How*

Google Works[145] puts it, "they possess business savvy, technical knowledge, creative energy, and a hands-on approach to getting things done." They work hard, but they want to impact you, and you cannot tell them how to think.

Gen Zs want to have meaningful jobs and be true to themselves, and they want the companies they work for to respect that. These young people want to drive culture change; they are now known as culture creators, which is why culture matters are growing in importance in companies. Gen Zs prefer to work in industries that they interact with in their personal lives as opposed to industries in which they don't frequently consume. This is why many leave companies early in their careers if they don't feel like their values fit in. For instance, many organizations in the oil and gas industry are seeing how their younger team members are migrating to other industries.

Gen Zs also prefer stable employment and individual tasks over team-based activities, but still, they value connection with a team. They know what they want and will go for it, as you can see with Greta Thunberg, the environmental activist.

Generational differences do exist and will continue to exist until the end of time. Every employee needs to be aware of these differences and understand how they can contribute to make the team better. While older generations can bring more experience, their perspective can be too conservative or outdated. Younger generations can bring new technologies and a fresh perspective, free of the bias of experience ("we have always done it that way") but can also lack the knowledge. Working together and implementing programs such as reverse mentoring can unlock the strengths of all of them.

Gender Bias

When growing their careers, women have always had to overcome unique challenges compared to men, particularly in science, technology, engineering, and mathematics occupations or other male-dominated jobs.[146] The COVID-19 crisis has only exacerbated the problems many women face in the workplace. Since the pandemic began, more women have reported that the working environment has become even more negative, challenging, and hostile.[147] How can women identify and address these obstacles to advance in their careers? What can organizations do to break down these barriers?

Statistics are unequivocal on the gender gap in corporate America. A recent study conducted by McKinsey and LeanIn.org[148] shows that for every 100 men promoted to manager before the pandemic started, only 85 women were promoted. And this gap was even larger for some women: Women held just 38% of manager positions, while men held 62%. The pandemic only intensified the challenges women were already facing in the workplace, such as:

1. **Work-life balance:** More women than men juggle a "double shift" — a full day of work followed by hours spent caring for children and household labor proved by decades of research as per the McKinsey report. Mothers also face persistent bias in the workplace. There's also a false perception that mothers are less committed than fathers and women without children. For example, when mothers take advantage of flexible work options, that perception is strengthened, even if they are just as productive as other employees.

2. **Higher performance standards:** Women are often held to higher performance standards than men, and they may be more likely to take the blame for failure, especially during times of crisis.

3. **Peer pressure:** The pressure of being the only woman in a group of men can be overwhelming: Senior-level women are nearly twice as likely as women overall to be "onlys" the only or one of the only women in the room at work. They are more likely to feel pressure to work more and experience microaggressions or harsh comments regarding their appearance, hair, or clothing, or even their competence.

4. **Lack of support and sponsorship:** Women are less likely to say their manager advocates for new opportunities for them, especially if there aren't other women to support them (if they are "the only ones" on a team of men). Women also tend to have fewer interactions with senior leaders. In the end, they end up not getting the network or the sponsorship they need to advance. When a female leader leaves one organization for another, it's not only a concern for the other female employees at the first

organization, but it's also an issue at the corporate level that must be resolved. Research shows that organizations are 50% more likely to outperform their peers when more women are at the top.[149] Women at the top also can help other women within the organization grow and help bring cultural change and diversity to the table.

5. **Industry bias:** Women remain highly underrepresented in software engineering; only three out of ten workers in the field of STEM in Latin America and the Caribbean are women.

How the COVID-19 Pandemic Impacted Women in Particular

The McKinsey research[150] conducted in partnership with LeanIn. Org shows that one in three mothers have considered leaving the workforce or downshifting their careers because of the COVID-19 pandemic. As per the research, "Mothers are more than three times as likely as fathers to be responsible for most of the housework and caregiving during the pandemic. In fact, they're 1.5 times more likely than fathers to be spending an additional three or more hours per day on housework and childcare." For the one in five mothers who don't live with a spouse or partner, the challenges are even greater, and financial insecurity is one of their top concerns.

The research shows that women are having a worse experience than men, but they are all facing other challenges. Black women, Latinas, Asian women, LGBTQ+ women, and women with disabilities are facing distinct challenges. They deal with more day-to-day bias in their workplaces, such as microaggressions, having their judgment questioned, or hearing demeaning remarks about themselves.

Indeed surveyed 1,000 U.S. workers[151] about their job sentiment this past year and expectations for 2021, and showed a similar negative impact. Men were significantly more optimistic than women about:

- Next year's job market (53% of men vs. 31% of women)

- Career opportunities (52% of men vs. 31% of women)

- Salary increases (47% of men vs. 29% of women)

- Work-life balance (54% of men vs. 38% of women)

- Productivity (60% of men vs. 45% of women)

The financial consequences of losing women in the workplace could be significant. The McKinsey research[152] shows that company profits and share performance can be close to 50% higher when women are well represented at the top.

Companies will have to dedicate more time to reshaping the workplace to address these issues and avoid losing more women.

Creating and Sustaining an Inclusive and Supportive Environment for LGBTQ+ Employees

Women are not the "onlys" in gender bias; challenges persist for LGBTQ+ employees who are lesbian, gay, bisexual, transgender, queer, and other sexual identities. There are many awareness periods to help stop the discrimination against LGBTQ+ people, such as May 17, the International Day Against Homophobia, Transphobia and Biphobia; May 19, Agender Pride Day; and May 22, Harvey Milk Day.

Still, corporations seem to be falling short in creating a safe environment. Research[153] shows that three in 20 LGBTQ+ women believe their sexual orientation will negatively affect their career advancement at work. For LGBTQ+ men, this number is even higher, at six in 20.

- LGBTQ+ employees experience "onlyness." LGBTQ+ women of color are eight times more likely than straight white men to report onlyness.

- LGBTQ+ women, especially bisexual ones, also experience more microaggressions, such as hearing demeaning remarks about them or people like them.

- They feel as though they need to provide more evidence of their competence.

- LGBTQ+ women are also more than twice as likely as straight women to feel as though they cannot talk about themselves or their life outside work.

- LGBTQ+ employees who don't feel safe enough at work won't self-identify, making them feel less happy with their careers and more prone to change jobs.

A report performed by Glassdoor[154] shows that LGBTQ+ employees are less satisfied at work compared to their non-LGBTQ+ counterparts. LGBTQ+ employees gave their companies an average overall company rating of 3.27 stars out of 5—that's below the average overall rating for non-LGBTQ+ employees (3.47).

Scott Dobroski, vice president of corporate communications and a member of Glassdoor's LGBTQ+ employee resource group, said "While many companies will turn their logos and social profiles to rainbows for Pride Month, creating a more equitable company is more than just symbolic or superficial moves. It's about action. Company leaders should take time to solicit feedback from their LGBTQ+ employees to better understand what's working well and what needs improvement to further support their workers."

As per Glassdoor research, among 10 employers with at least 25 reviews by LGBTQ+ employees:

- Apple has the highest overall company rating among LGBTQ+ employees with a 4.14 rating, followed by Starbucks (3.56) and Target (3.31).

- Wells Fargo has the lowest overall company rating with a 2.65 rating. Other lower-rated companies include Walmart (2.70) and Amazon (2.85).

How can companies create a positive and safe working environment for all employees, regardless of sexual orientation or gender identity and expression?

I interviewed Lin Cherry, chief legal officer and head of diversity and inclusion at Wizeline, a software development and design services company with 1,100 employees, who said, "If people don't self-identify, they are not living their life to the fullest."

Watch the interview with Lin Cherry at
https://youtu.be/t5J9Y-8CCGE.

She recommends that companies demonstrate their support for the community visibly. It is not easy for all companies to create an authentic, inclusive culture. Wizeline, for example, had a very diverse employee base from the beginning, given that the founder,

Bismarck Lepe, was a minority himself, being a son of immigrants from Jalisco, Mexico. The company has employee resources groups (ERGs) that organize LGBTQ+ open talks such as "my journey as a gay man in tech" or "happy to be me a transgender woman." They "spark other Wizeliners to feel open about telling their story and promote self-identification."

Companies can start making a difference by following these specific recommendations to retain LGTBQ+ employees:

Create structural support for trans employees. This includes making health coverage inclusive of trans people, supporting leave for transitioning colleagues (including bathrooms with all-gender options), allowing changes to documents and records, and ensuring that HR systems are inclusive of all employees' genders and pronouns.

Many organizations and platforms, such as LinkedIn, allow employees or users to choose their preferred pronoun ("he/him," "she/her," or "they/them") or declare their gender. A person's gender identity is not restricted to being either a man or a woman. Some people do not identify with any gender, while others identify with multiple genders. For example, if you prepare a survey or a form, consider using the following options: male, female, or nonbinary.

A person's sex is typically based on certain biological factors, such as reproductive organs, genes, and hormones, but still is not binary, just like gender.[155] A person may have the genes that people may associate with being male or female, but their reproductive organs, genitals, or both may look different. Teammates can assume but can't for certain determine gender by voice or appearance. It is good for organizations to ask individuals how they express themselves to others and how they want their team to see them and how they would like to be addressed.

Provide training to all employees to prevent and address micro-aggressions and demeaning behavior, encourage a pronoun-friendly culture, and create reporting channels to investigate and correct inappropriate behavior.

Get certified by the Human Rights Campaign (HRC). Launched in 2002, the HRC foundation's corporate equality index has become a roadmap and benchmarking tool for U.S. businesses in the evolving field of lesbian, gay, bisexual, transgender, and queer equality in the workplace.

People With Disabilities, Caregivers, Other Personal Circumstances

Being a person with a disability, being sick, being a caretaker for someone who is ill, being a pregnant veteran or military spouse, or having mental health challenges are other examples of employees who may have different needs at work. One trillion dollars per year are spent globally for the loss of productivity because of poor mental health, including anxiety.[156]

There may be policies for women with children, but what about men who need to care for an elder? These are examples of "untapped talent pools."

Anyone can experience similar challenges at any moment in life. Just by getting sick, getting pregnant, or growing old, your circumstance may change. That is why Elsa Sjunneson, author of *Being Seen: One Deafblind Woman's Fight to End Ableism,* said in an interview,[157] "Start looking at it as a way of being. It is not binary, able, or disabled. You can find yourself disabled in a matter of moments."

She has been deafblind since birth. She highlights that she doesn't want nondisabled people to pretend disability doesn't exist; instead, she would like disabled people to be welcomed in the world and be received with warmth and accommodation instead of fear and rejection.

Businesses sometimes focus only on the legal obligations of providing accessibility support, but it is really challenging to meet everyone's needs. HR doesn't have to be aware of everything, but they should have honest conversations with the new hires and employees on what accommodations they need. For example, for some, working from home probably makes work easier. On the other hand, maybe an employee needs to bring a service dog, cannot read a whiteboard, or can't read small print. Some tools can help accommodate these people, such as glasses, ramps, or Bluetooth hearing aids, but these are only tools if we don't incorporate new behaviors. People need to be curious to find out what can help instead of being afraid to ask.

Elsa says that "90% of my day is explaining this is what I need, but it is not a burden. It is the idea that we are not welcome that is a burden." Acknowledging and accepting disabilities is part of helping people show up, so they don't hide any part of themselves.

The part they want to hide could make a difference in the business or even the world.

> *The mandate is to stop being so afraid of disability and start adapting to it. There needs to be change in our culture.*
>
> —Elsa Sjunneson

Mental health is also an issue that is, fortunately, getting more attention from organizations. In the past, mental health would not be considered a real problem, just maybe a "weakness." Some of the drivers or mental health issues are anxiety, stress, and depression.

Depression, for instance, as per the Mayo Clinic, is a mood disorder that causes a persistent feeling of sadness and loss of interest. Also called major depressive disorder or clinical depression, it affects how you feel, think, and behave and can lead to a variety of emotional and physical problems. You may have trouble doing normal day-to-day activities, and sometimes you may feel as if life isn't worth living.

Depression is not a weakness or something you can simply overcome like a to-do list item. It may require long-term treatment with specialists and therefore may require special attention from family members, coworkers, and teammates.

Some people may suffer from mental health only temporarily, during a certain circumstance, such as being less busy, being too busy, or having a change in the family. But others can have multiple episodes.

For instance, most people experience depression or anxiety when they lose a dear person. I lost a pregnancy and one of my siblings, and I can assure you I didn't get better in three days, which is the typical time off companies provide. Grief may take from six months to two years, or more. It is a process that needs to be normalized in companies, accepted, and adapted to every circumstance.

Pregnancies and adoptions are also special circumstances. Maternity leave in the United States is still too short compared to other countries. There is a mandate of six weeks of unpaid time off in some states, but most Americans do not have access to paid family leave through their employer, compared with other countries that have several weeks of paid leave, such as Britain (39 weeks) and Japan (52 weeks).[158] In Canada, there is a maximum of 15 weeks of maternal benefits, plus parental leave is offered to parents (no matter the gender) who are caring for a newborn or newly adopted child or children.[159]

Life returning to work after pregnancy or adoption is not easy either. Parents may require time for doctor's appointments, and mothers may require time and a special place designed for lactation or to pump milk for the newborn.

Some recommendations are:

- Provide more days off or even unlimited days off in all these circumstances (grief, depression, anxiety, caring for a loved one). Fortunately, when I lost the pregnancy, I was working for a company that allowed me to take the time I needed. Even though it is a long process, it is important that the person feels comfortable when going back to work. That timing could be very different for everyone.

- Be patient and listen to understand their needs

- Provide ongoing assistance; most of these processes are long or never end, so they need to be understood and followed up over time.

- Provide spaces for breastfeeding moms to pump milk, store it, rest or clean, and change their clothes, which may get wet during a long day of work.

- Train leaders to know how to react, understand what they can offer, and be flexible to accommodate their needs.

- Provide flexible schedules, four- or six-hour shifts, and night shifts.

Accommodating for different needs will improve the engagement and satisfaction of these employees and improve their quality of work and work-life balance, reduce stress and anxiety, and offer an alternative to just leaving the job.

How to Build Diverse Teams

It is time to see diversity as an opportunity, not an obstacle. Companies may not have all the answers, but employees can propose long-term solutions that enable them to feel safe and collaborate more if leaders are open.

Companies are faced with many barriers that prevent people from feeling equal. It's time to review the entire employee experience of the minority workforce and identify process weaknesses. Statements are what we say we do, but culture is what we do. Companies need to build DEI into their culture.

More innovation and resilience to change are possible in a company where employees' differences are respected and ideas can be expressed freely.

To retain minorities in the company, DEI must impact individual and team behaviors and must be embedded in the entire employee experience: recruiting, hiring, onboarding, engaging, performing, developing, and even in the departing process.

We Culture Behavior: Empowering Employees to Overcome These Challenges

Many organizations have taken important steps to support employees, especially during the COVID-19 pandemic. They are sharing valuable information with employees, including updates on the business's financial situation and details about paid leave policies. To aid in overcoming the challenges, organizations should consider following these steps:

1. Acknowledge There Is a Problem

You may think this is not happening in your company. Or maybe you have worked hard to improve diversity, but employees don't see any changes in behaviors. Marc Benioff, CEO of Salesforce, in his book *Trailblazer*,[160] talks about how he tried hard to beat bias in Salesforce, but over time the gap persisted. In 2016, only 2% of Salesforce's employees were African-American. He says, "Equality is a moving target" as everyone has unconscious bias. You need to acknowledge the imbalance and work continuously to address it.

As a first step, we always urge companies to acknowledge the difference. Differences do exist and will continue to exist. We cannot afford to act like this is not happening. Every employee needs to be aware of these differences and understand how, no matter what, they can still all work together as a team.

2. Listen and Speak About It

Leaders need to open themselves, listen to employees deeply, learn from them, and avoid being defensive to make corrections.

Internal communication and external communication to clients, suppliers, and even the community are essential to the bottom-up approach. As different generations communicate differently, communication methods also need to be diverse. You may need to use social media for millennials, apps with images and pictures for Gen Z, manuals and internal newsletters for Gen X, phone calls for baby boomers, and face-to-face communication for everyone as much as possible. Gossip and informal communication are becoming more powerful, keeping secrets no longer safe. The more decisions and results are communicated, whether formally or informally, the better.

Communication has to go both ways, so be ready to receive feed-back. Sometimes you may not like what you hear, but you still can do something about it. An MIT[161] paper shows that "Cultivating and maintaining ties across functional and geographic lines was the largest network-related predictor of advancement and retention." Organizations should promote connections to senior people in the hierarchy and to similar peers through employee resource groups.

3. Take Action

You can have a diverse team, but that doesn't mean the members feel psychologically safe to speak up and give their best. You have to build a system that supports and promotes the behaviors required to lead a diverse workforce. Sally Helgesen, cited in *Forbes* as the world's premier expert on women's leadership, highlights that inclusive behaviors must be demonstrated at the leadership level or they will have no real impact in the book, *Leadership in a Time of Crisis*.[162]

Among the best strategies to overcome the challenges of being underrepresented are empowering them through:

Being part of minority networks and support groups (formally or informally). Many organizations are forming equality groups so like-minded people can openly discuss issues that concern them. Self-employed women on their side can certify their businesses as women-owned and receive the benefits of being part of a network of women, such as Women's Business Enterprise National Council,[163] the International Association of Women,[164] or the U.S. Women's Chamber of Commerce. Minorities also can form their own small network. For example, when I moved to Beaumont, Texas, it was a male-dominated industrial town with barely any Latinas participating in groups or networking events. So I teamed up with other Latinas business owners, and we founded the SETX Hispanic Women's Network to help empower women in the area who speak Spanish and want to grow professionally.

Identifying fears or barriers that hold them back. From working on gaining self-confidence to teaming up, women can achieve a better work-life balance. In many cases, women feel they must be in charge of most of the household chores, or at least that is the tacit agreement they have with their partners. Women must start having the conversation about sharing the burden and working in teams with other family members. The same happens at work: Women tend not to work in teams, sometimes so they don't look weak.

Getting a mentor or sponsor within the organization or an executive coach. A lack of mentors and advisors can stunt your professional growth. These fears and barriers can be unconscious, so getting a coach or mentor can help identify strengths and improvement opportunities, increase self-confidence, and become an accountability partner to move — slowly but steadily — in the right direction.

Playing to their strengths is key, as is accepting that bias continues to exist. Yes, minorities face unfair biases in the workplace, but that doesn't have to stop them from trying to grow, showing their strengths, and helping others grow too. Minorities bring unique perspectives, ideas, and experiences to the table. For instance, the McKinsey and LeanIn.org research shows that senior-level women have a vast impact on a company's culture:

"They are more likely than senior-level men to embrace employee-friendly policies and programs and to champion racial and gender diversity. Over 50% of senior-level women say they consistently take a public stand for gender and racial equity at work, compared to roughly 40% of senior-level men. And they're more likely to mentor and sponsor other women: 38% of senior-level, compared to only 23% of senior-level men."[165]

Improving the Employee Experience to Increase Engagement

You can have a diverse team, but that doesn't mean the members feel psychologically safe to speak up and give their best. You must build a system that supports and promotes the behaviors required to lead a diverse workforce. The real benefits start when new behaviors are learned and reinforced continuously, and the organization's culture changes to prioritize respect over peer competition.

Engaging employees is about making them communicate and collaborate as much as possible, but especially about caring for them and making them feel unique and needed from the very first minute they join the company until they leave.

Engagement impacts all the employee interactions with an employer, from recruitment to departure, known as the "employee experience." Every step plays a role in how a worker feels about an employer.

Employees' willingness to stay at the company, collaborate, and recommend it to others, or disconnect and leave, are based on the employee experience. Companies need to ensure the culture, the brand, and the purpose (what the company does, communicates, and desires) are aligned with the employee experience and displayed every step of the way.

The same way a retail store is structured and designed to capture customers' attention and provide the best experience to get the sale, workplaces should be like this for their employees. Especially when challenges arise, such as the development of hybrid teams or the increase of turnover rates, the employee journey needs to be reviewed to ensure the employee experience is still consistent, engaging, and rewarding for all team members.

While the leaders' work has a huge opportunity to impact the employee experience, a great employee journey will help guide them. According to Deloitte's Human Capital article,[166] "Governing Workforce Strategies," there are some questions to ask that can translate values into action:

- **Workforce social contract:** How does your organization treat its employees, contractors, and service providers of every type?

- **Meaningful diversity:** Are workers from diverse communities in a position to wield influence in the organization?

- **Human capital brand:** How are your culture, workforce, and leadership being portrayed externally?
- **Culture risk sensing:** What signals are you seeing that point to outliers in worker behaviors and norms?

Helping the workforce make the shift to your company's culture can help them in their own journey.

What Are the Steps of the Employee Journey?

How you hire, onboard, or engage your people can help shape the culture, reinforce it, or push it in a new direction. Unless the people fit the culture, it's hard to maintain the company's values over time. That's why the employee experience should not be solely HR's job; it is everyone's job in the organization.

It is possible to create a team that demonstrates the desired behaviors on a consistent basis. Review what you are doing across the different steps of the employee journey (see Figure C9.1) and make sure to communicate and apply the 12 CARE skills across the entire 7-step employee journey.

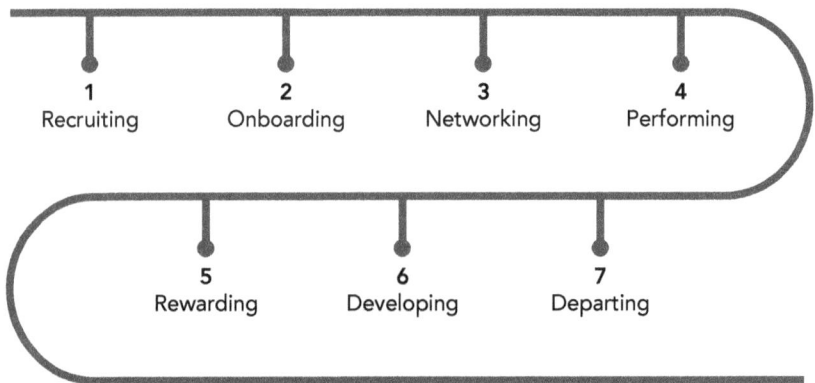

1 Recruiting **2** Onboarding **3** Networking **4** Performing

5 Rewarding **6** Developing **7** Departing

Figure C9.1   Seven steps on the workforce's journey.

Recruiting

How and where are you looking for talent? Are you using websites or LinkedIn? Are you interviewing candidates at university job fairs? Where you recruit will tell a lot about the types of candidates you are going to have. If you are looking for a more diverse pool, you may want to change the way you recruit or the image candidates have of your company. Some companies are encouraging employees to reach out to their network to find potential candidates. One of the characteristics of these new generations is that they talk about brands with family and friends, so people they know will be more inclined to accept an offer from a company recommended by a friend.

What do you have to offer? The best talents want to work for organizations that have a strong company culture and live by it. To recruit new generations in particular, it is important to show them the benefits of belonging to the organization and not just focus on the salary. They are usually interested in being part of organizations that show interest in the community and the environment. They also want to know about career opportunities, training possibilities, flexible hours, and vacation time.

Organizations like Glassdoor publish company reviews about CEO's leadership, insights into jobs, and work environment reviews based on the input of employees who voluntarily provide anonymous feedback. Blind is another source of information, where employees follow discussions on company perks, salaries, and management style shared anonymously before accepting an offer.

Millennials are rethinking their work-life balance, especially after the pandemic, and want to take advantage of remote work policies, so the possibility to work remotely could be very important for many candidates. The way to get the candidate to apply also counts. Some companies like McDonald's and Paradox[167] are starting the process through text messages or using voice applications through Alexa or Siri.

Companies should be recruiting all the time and everywhere, regardless of their current needs. If remote work is an option, the pool of candidates is far greater across the country or even across the globe.

Review how your company is evaluated and what other companies are offering. Company image is the first step in recruiting the best talent, so showcase your strengths in every interaction with clients and potential candidates. A Gallup report also shows that engaged employees are 23 times more likely than disengaged employees to strongly agree that they would recommend their organization as a great place to work, impacting your organization's brand reputation.[168]

The interview process can be very time-consuming both for employers and candidates. The more you streamline the process, the better for everyone involved.

Put yourself in the shoes of job candidates who have to answer the same questions again and again for every opportunity. Companies that make the process easier can get more candidates, especially those who may not be looking for a job and could be great future employees. Reduce the process by making it more automated, taking advantage of what is already online on LinkedIn or on a resume, and making applicants answer only the five questions that will indicate the job requirements, such as years of experience or certifications. Simply, these are "go/no go." The process will be much faster and reduce the application time from 20 minutes to two minutes.

Automating the process to schedule a quick interview or respond to the candidate for a yes or no can also be a great way to improve the candidate experience. Remote interviews can make the first interviews a lot easier for candidates.

The cons of standardizing the first filter could be missing great talents because they don't have the exact experience you are looking for, but could have experience in consulting or other areas that bring valuable knowledge to the table.

Hiring is the stage to clarify the company's values. This is for the employee's sake as well, to evaluate a good cultural and technical fit. It also prevents the company from spending money on someone who will be disengaged and leave soon or will not perform based on the same values.

The interviewing model is another step that should be analyzed. Some companies hire for skills, and others hire for specific jobs. Hiring for skills may allow companies to access a broader pool of candidates.

The company Ideo likes hiring Unicorns, people who don't necessarily fit in a prescribed role but have something unique, a perspective or a set of skills. Others have two interviews, one for technical fit and one for culture fit; both need to be a "yes." Culture fit is key to keeping the employee engaged for a longer period.

While in the past the candidate had to dress for success and mount a show to capture the attention of the recruiter, in a candidate market, interviewers also need to be careful about how they ask questions, how they react, and how they communicate the employee value proposition to capture the candidate's attention.

Review the technology used for recruiting to ensure it is not biased. Use a variety of recruiting sources if you are not able to access or capture minorities' attention.

Reduce biases during the hiring process. Sometimes people don't even take a job when they feel the company is not inclusive or safe. Training recruiters by providing recruiting booklets and best practices, and implementing blind resume screenings (removing names and gender signifiers) can also help reduce unconscious bias during hiring.

Ask diversity-related questions at the application stage. This is challenging. You don't want applicants to think their cultural background, gender, or physical ability could impact their application. The sharing of diversity information should be completely voluntary. Frame those questions correctly, be transparent about your goals and why the question is important, and highlight that answering is voluntary. Instead of asking too many questions, you may want to ask specifically if they have certain needs to be more comfortable at work, especially for employees with disabilities.

Define a clear employee value proposition and train recruiters to talk about it. Some 65% of candidates report they have actually discontinued a hiring process due to an unattractive employee value proposition.[169]

Onboarding

New hire training or onboarding is very important in We Culture companies, especially for employees who are hired to work remotely. Companies used to "teach" their culture to new employees simply by sharing spaces and attending meetings. That is how employees were supposed to understand the rules of engagement and performance organically. Onboarding was supposed to be mostly an administrative process. But it is much more than that. While senior employees learned a culture onsite, remote hires won't have that same onboarding experience. The process should improve their networking from the start, reduce their learning curve, and ensure a cultural fit.

Zappos, for example, provides four weeks of customer-service training to everyone who joins the company. It offers $2,000 to those brand-new employees who want to leave the company right after the onboarding, as proof of loyalty, culture fit, and long-term thinking.[170]

The onboarding needs to drive the employee to understand their specific role and how to interact within the organization. As the first connections within the company are usually the most crucial for development and engagement during the first months, some managers now engage newly hired individuals in a series of activities designed to cultivate key connections in their first three to six months.

The other challenge companies have to deal with when working remotely is training new hires or junior employees on technical matters. Globant, for instance, developed "augmented coding," a tool to help junior coders get technical suggestions. While it cannot replace leaders, it ramps up their productivity and reduces mistakes using artificial intelligence.

Successful onboarding experiences can create positive memories that will shape employee connections (connection dimension), engagement, and performance in the future.

Welcome employees. Employee resource groups (ERGs) can help new employees to get comfortable during the onboarding process.

Help employees get to know themselves by doing personality tests. These tests will help them understand where they are compared to the average team member and the company's desired behaviors.

Networking

This stage usually involves the employees. How are they feeling at work? Are they enjoying it? Are they bringing their whole selves to work, that is, their strengths and abilities? As reviewed in previous chapters, with the engagement rate at 33% in most companies in the United States vs. 70% in the world's best organizations, this stage of the employee journey needs to be examined in detail. Gen Zs, for instance, want to seek meaningful jobs and be true to themselves — that's why many leave companies early on in their careers if they don't feel their values fit in.

Companies need to look for ways to improve how employees can work at their best while feeling respected and appreciated.

Creating open meetings with no agenda, face-to-face or online, where everybody can jump in and connect, doing all-hands meetings as well as individual one-on-ones and coaching sessions, organizing formal and informal events, launching projects — all these opportunities to connect should reinforce the 7Rs of the company (see Chapter 1). You can even share real stories about people who have demonstrated those values or training an internal team of employees to become culture coaches and ambassadors.

Communicating clear values is even more beneficial when companies grow too fast to help align employees' behavioral patterns. Sharing values is not just about including them in the onboarding package or posting them on the wall. It is about living, working, and breathing based on these values. Leaders, in particular, are the role models for these values. They have to *make the culture come alive regularly,* and walk the talk, even in a remote conversation, and reward people who are role models for those values.

Implementing synchronous and asynchronous communications is always vital to ensure employees have different ways to communicate based on their various needs.

We cultures can help drive engagement through peer support and a collaborative environment where everyone feels safe and appreciated, and engagement doesn't depend on a single person like an immediate supervisor or a department like HR.

Create ERGs for each underrepresented group. ERGs are voluntary, employee-led groups made up of individuals who join together based on common interests, backgrounds, or demographic factors such as gender, race, or ethnicity. ERGs bring people together to learn about each other and promote respect for everyone no matter what. They create a safe space and organize activities that are open to everyone, not just people of the community, such as films or talks, to learn from and support each other. They make sure employees have an opportunity to be heard, valued, and engaged.

Provide training to all employees to prevent and address micro-aggressions and demeaning behavior. Encourage a pronoun-friendly culture and create reporting channels to investigate and correct inappropriate behavior.

Facilitate feedback and open more personal conversations. You can do this through gratitude sharing, virtual happy hours, meditation and mindfulness, or Slack channels to discuss specific topics. Deploy alternative complaint systems[171] to avoid retaliation, or workers who complain will end up facing career challenges or experiencing worse mental and physical health. Complaints should not be seen as threats but as feedback to drive organizational change.

Create buddy programs. Randomly pair people to get to know each other. You can do peer or group coaching with other team members or even other departments.

Promote and support events and organizations related to issues in your organization. Examples of this include the Didi Hirsch marathon for mental health and suicide prevention.

Offer hybrid work. Remote is an opportunity for existing employees and women—particularly mothers, caregivers, and people with disabilities—to return to work while maintaining work-life balance. Organizations should explore the possibility of offering remote work opportunities as a benefit, or at least provide a hybrid workplace with the option to work from home two to three days a week.

Promote inclusivity in remote-working environments. Working from home and videoconferencing can expose more minority communities because others can see their personal lives and make them feel more isolated. Leaders and team members should help by offering more frequent one-on-one sessions to see how they are doing, make sure everyone participates in meetings, and be more aware of personal needs.

Put them in front of local networks. Research by MIT[172] shows employees who experienced rapid promotion and those who stayed with the organization longer were the ones who were able to develop ties with key stakeholders within the organization and with local networks. One black male employee helped people trust him: "People want to know they can give you a problem and be safe to take it off their plate. The more you do this for people and reliably produce, the more you get pulled into bigger and bigger opportunities."

Performing

This stage includes how employees perform and what companies do to help them perform better. In Part 2 of this book, on the attention dimension, we discussed different ways to measure performance that will help improve engagement in this stage.

Develop and track metrics. By collecting and analyzing data on diversity over time, companies can increase accountability and transparency around diversity issues, such as if the organization has a lower representation of women in managerial positions relative to the local labor market, similar firms, or the goals of the corporation. This identified shortfall can lead to concrete goal-setting about numbers and timelines for increasing the representation of women in management.

You can also measure employees as part of the board of directors, senior positions, or managerial positions, or measure the attrition of the diverse population; if they don't feel comfortable, they tend to move jobs more often. You can also get more subjective through input and surveys from your employees. If employees don't feel comfortable with this, you can do it through anonymous pulses (surveys), for instance.

It is essential to continuously track outcomes for promotions, raises, and layoffs by gender. Bias is not easy to remove. Sometimes you may achieve improvements one year, but go back to square one the next year.

Use surveys to identify high-risk groups, target communication, and show you care. Pulse surveys can identify teams, departments, or locations that are seeing increased levels of stress, anxiety, or burnout. These surveys provide instant information, as they are short and easy to analyze and can be sent regularly (typically only 5-15 questions; therefore, they are not time-consuming and are very easy to answer). They can also provide useful information for the employees, giving them a sense that someone is taking care of them and can provide an answer as to why performance is decreasing.

Rewarding

This stage includes how employees feel and are rewarded for their accomplishments. Chapter 5 explains many different ways of rewarding employees. It needs to be a fair system that reflects diversity commitments. If the reward system is not fair, all other efforts will be fruitless.

The ROI of taking care of employees can show that boosting the lowest salaries and reducing the highest could be a good place to start in addressing compensation inequality.

CEO's earnings are usually 20 to 300 times more than the median of employees they work with, while some trailblazing companies reduce this gap to increase engagement. For example, companies are trying with peer-based salaries, self-set salaries, cutting dividends, eliminating operators' hourly wages, and offering fixed salaries, especially for salespeople.

A good example of this change is Gravity Payments CEO Dan Price, who raised the minimum wage[173] to $70,000 annually and cut his own salary from $1.1 million to $70,000 to help fund it. Six years later, in August 2021, Dan proved his perspective. He reported that revenue had tripled, the customer base had doubled, 70% of employees had paid down debt, many bought homes for the first time, 401k contributions grew by 155%, and turnover dropped in half. His business is now a Harvard Business School case study. Dan tweeted "6 years ago today I raised my company's min wage to $70k. Fox News called me a socialist whose employees would be on bread lines. Since then our revenue tripled, we're a Harvard Business School case study and our employees had a 10x boom in homes bought. Always invest in people."

Key Insight
Invest in people first.

Foster inclusivity through pay parity. An organization has to audit its pay structures to maintain parity consistently. Salary is a hard measure for inclusivity, but it is not easy to maintain. It has to be monitored continuously.

Developing

Companies need to find new ways to continue developing their employees. Development is one of the things new generations appreciate the most at work.

Paying tuition or delivering training are the most basic ways to help with employee development. Other ways are offering one-on-one coaching sessions through leaders or peers, as discussed in Skill #9. You can provide formal mentoring programs with external coaches. Most programs start with an executive coach meeting with the CEO.

Making employees participate in their own development is a characteristic of a we culture. Companies can use more resources from within the organization to succeed in employee development. Some employees may be natural leaders, emotional supports, peer coaches, or experts in yoga or mindfulness sessions. Organizations need to identify potential employee volunteers to be trained as advocates or ambassadors who can organically connect better with other employees to dive results or engage them.

Upskill leaders to have caring conversations. Employees learn organizational culture not by merely reading a guide but by observing leaders, and seeing what is rewarded and what is penalized. Leaders of all levels, genders, and races need more training and guidance on how to best support women and minorities; deal with unconscious bias, diversity, and inclusion; handle difficult conversations; and bring more vulnerability to work. When managers start by opening themselves to their issues and challenges, employees gain more trust in them and feel more comfortable being open.

Leaders and middle managers should have conversations with their employees more often than when working face-to-face, and not all of them should be focused simply on "did you meet your objectives." They should also be comfortable driving informal conversations about general well-being. If this is not something they are used to doing, leaders should be trained on soft skills such as empathy, bias, mental health, psychological safety, and dealing with difficult conversations. During one-on-ones, leaders can even start by asking simple questions to make sure employees sleep enough, eat well, exercise, and have time off as needed.

As a leader, be vulnerable about your struggles. Leaders should be more candid and open about how they feel and how they also struggle. If they are the first to share that they are using mental health services or other well-being resources, it sends a message to the employees that self-care is OK and expected.

Not everyone can lead with issues and crises the same way. You may need an external coach yourself to help you deal with your own struggles. A coach should give you tools to get better and be more prepared to help others.

Provide one-on-one coaching opportunities. Provide sponsorship to support career progression and individual coaching sessions for minorities.

Suppose women cannot get support and sponsorship organically. In that case, organizations can bring in external coaches to help bridge the gap and provide the extra support needed for women to feel empowered to move up the career ladder. Ning Wang, CEO of Offensive Security, a cybersecurity training and certification company based in New York, said in an interview[174] that she hired an executive coach for herself and the executive team "to define what kind of team we want, what kind of norms we want, and the behaviors we want to have." She added, "We really want to bring that way of working and being to how we interact daily. The coach first worked with the entire executive team and me. Then the next year, she worked with the next level of management, our directors, and managers. And the next year, we had her work with the whole company."

Provide mental health counseling. Corporate well-being and mental health support are no longer fancy perks but must-haves. Organizations are offering services related to mental health, such as counseling programs, and they are providing tools and resources to help employees work remotely. "Mental health has impacted us in ways we haven't seen before, the global workforce as a whole is currently dealing with a lot of mental health issues," said Divya Ghatak, chief people officer at SentinelOne, a cybersecurity startup based in Mountain View, California. For instance, the company brought in mental health experts; offered yoga, mindfulness, and meditation classes; and offered a mental health day off once or twice a quarter.[175]

Expand employee assistance program offerings. Large organizations offer employee assistance programs. Remind employees that they are available to address serious concerns, such as depression and anxiety. If your company is not offering these programs, look for other ways to provide other services that contribute to mental-health assistance, such as individual coaching, yoga classes, and meditation.

Use surveys to identify high-risk groups, target communication, and show you care. Pulse surveys can identify teams, departments, or locations that are seeing increased levels of stress, anxiety, or burnout. These surveys provide instant information, as they are short and easy to analyze and can be sent regularly (typically only 5-15 questions; therefore, they are not time consuming and are very easy to answer). They can also provide useful information for the employees and give them a sense that someone is taking care of them.

Use the strengths of individual employees to share resources internally. Companies are starting to use more resources from within the organization. Some employees may be natural leaders, emotional supports, peer coaches, or experts in yoga or mindfulness sessions. Organizations need to identify potential employee volunteers to be trained "well-being advocates" who can organically connect better with other employees.

Offer flexibility. Through getting better-quality feedback, you may understand that, for instance, offering three months of maternity or paternity leave is just a short-term benefit that doesn't address the holistic needs of the employee. When new parents come back to work, they may need a more flexible schedule, or they may feel they cannot work as much as before and, therefore, their growth opportunities will suffer.

Identify the differences that matter to employees. Rather than just assuming that "woman" or "LGBTQ+" is the identity that matters most, organizations need to address other factors that may be more important. The person may be a caregiver or may have mental health issues.

Recognize and reduce microaggressions. All employees need to be trained to acknowledge what microaggressions are and the impact they have. They need to realize that behind any comment or behavior, the intention is not to hurt the other person.

Offer consistent training on unconscious bias and diversity. Workshops and role-plays can help employees understand better what makes certain people uncomfortable or offended. A one-time training is not enough, as some behaviors are hard to identify and adjust.

Help in their development. Provide sponsors to support career progression and individual coaching sessions.

Eliminate onlyness. Banishing onlyness not only replaces the goal of gender parity, but it also will diminish some of the barriers that minorities have to deal with. For instance, instead of having just one Latina in your board, make sure you have at least two.

Provide mentors. The same MIT research shows that people of color with higher promotion and retention rates were more likely to have sought and received mentoring and career advice from others in their networks. These networks included individuals in other units, geographies, or roles; some were peers or colleagues who were at most one level up in the organization.

Involve the ERGs in development. ERGs help build a workforce that reflects the demographics of their customer base and provide ideas on increasing the organization's spend with diverse suppliers. They can also be great partners for identifying gaps in an organization's talent development process.

Departing

Finally, there may come a time for the employee to leave the company. The employee may decide to leave voluntarily, or the company may determine that he or she is no longer a good fit for the company, but the relationship has a closing. We Culture companies rarely decide to dismiss an employee, as usually the system makes the employee feel unfit and decide to leave on their own. An intentional culture sustained throughout the entire employee experience doesn't allow people to feel comfortable even when they don't fit. Simply because everyone else fits so well, it is hard for them not to follow the rest. When cultures are not that intentional, undesired behaviors can remain for years without being challenged.

Still, surveys can help provide information about why employees choose to leave the company to ensure the reason is unfitness and not an employee experience issue that needs to be fixed.

We organizations can even decide to keep, intentionally, a connection with the departing employee, in case he or she may be rehired for a future job, provide consulting, or refer other candidates. This is a way to respect even the people who decide to leave at one moment in time but as part of the family may be welcomed back again. This is part of the growth mindset. Never close a door.

After crafting a program considering all these recommendations, continuously review its effectiveness. A survey showed that even though companies are putting in more overtime work, employee engagement has not increased since 2016. For instance, 87% of employees have access to mental and emotional well-being offerings, but only 23% use them. And 48% of employees who utilize well-being programs report being highly engaged, compared to 30% of employees who do not. Gartner[176] recommends three strategies to boost employee participation in offered programs: 1) increase employee understanding of well-being needs and offerings; 2) reduce well-being stigma; and 3) reduce the time and effort needed to participate in well-being programs.

It's important to reflect on organizational customs, rituals, and norms to ensure they're inclusive, specifically regarding flexibility and psychological safety.

Hands-on 9.1

Developing Diverse Teams

Within your team, develop the best-selling product ever:

- Estimate a limited time (no more than 15 minutes)
- Choose four team members to be observers
- Assign the rest of the team members the responsibilities of:
  - Time keeper
  - Designer
  - Engineer/buyer
  - Motivation specialist to choose rewards
  - Other positions you may require
- After the exercise, the observers should report what they noticed in regard to the different skills used by each function, and whether the differences helped to accomplish the objective.
- What went wrong? What should change in the future?

Summary

Increase employee engagement by respecting everyone and everything.

Recap ①
Build empathetic one-on-one relationships.

Recap ②
Promote team psychological safety.

Recap ③
Embrace diversity, equity and inclusion.

Reflection Time

Take five minutes to think about three highlights from the dimension of RESPECT. Write them on your note pad or the action plan available on the We Culture app.

Three Highlights

_____

_____

_____

References

128. McKinsey & Company, "All In: From Recovery to Agility at Spark New Zealand," *McKinsey*, McKinsey Quarterly, 2019, https://www.mckinsey.com/industries/technology-media-and-telecommunications/our-insights/all-in-from-recovery-to-agility-at-spark-new-zealand.

129. Rocío Lorenzo, Nicole Voigt, Miki Tsusaka, Matt Krentz, and Katie Abouzahr, "How Diverse Leadership Teams Boost Innovation," BCG, 2018, https://www.bcg.com/publications/2018/how-diverse-leadership-teams-boost-innovation.

130. Catalyst, "Women CEOs of the S&P 500 (List)," *Catalyst*, 2021, https://www.catalyst.org/research/women-ceos-of-the-sp-500/.

131. Catalyst, "Women CEOs of the S&P 500 (List)," *Catalyst*, 2021, https://www.catalyst.org/research/women-ceos-of-the-sp-500/.

132. MIT Management Executive Education, "Special Collection, Tools for Change: Advancing Equality in the Enterprise," *MIT Sloan Management Review*, Summer 2021.

133. Vivian Hunt, Dennis Layton, and Sara Prince, "Why Diversity Matters," *McKinsey*, 2015, https://www.mckinsey.com/business-functions/people-and-organizational-performance/our-insights/why-diversity-matters.

134. Rocío Lorenzo, Nicole Voigt, Miki Tsusaka, Matt Krentz, and Katie Abouzahr, "How Diverse Leadership Teams Boost Innovation," BCG, 2018, https://www.bcg.com/publications/2018/how-diverse-leadership-teams-boost-innovation.

135. Marc Benioff, and Monica Langley, *Trailblazer: The Power of Business as the Greatest Platform for Change* (New York: Penguin Random House, 2019).

136. "The Power of 'Out' 2.0: LGBT in the Workplace," *Center for Talent Innovation*, February 1, 2013.

137. Bayard Love and Deena Hayes-Greene, "The Groundwater Approach," Racial Equity Institute, https://www.racialequityinstitute.com/groundwaterapproach.

138. McKinsey & Company, "Race in the Workplace: The Black Experience in the US Private Sector," *McKinsey*, 2021, https://www.mckinsey.com/featured-insights/diversity-and-inclusion/race-in-the-workplace-the-black-experience-in-the-us-private-sector.

139. Kevin Sneader and Lareina Yee, "One is the Loneliest Number," *McKinsey Quarterly*, January 2019, https://www.mckinsey.com/featured-insights/gender-equality/one-is-the-loneliest-number.

140. Briana Krueger, "Why are Millennials Leaving Their New Jobs? The Reasons Might Surprise You," *Delivering Happiness*, https://blog.deliveringhappiness.com/why-are-millennials-leaving-their-new-jobs-the-reasons-might-surprise-you.

141. https://www.gallup.com/workplace/321965/employee-engagement-reverts-back-pre-covid-levels.aspx.

142. D'Vera Cohn and Paul Taylor, "Baby Boomers Approach 65 – Glumly," Pew Research Center, 2010, https://www.pewresearch.org/social-trends/2010/12/20/baby-boomers-approach-65-glumly/.

143. CareerBuilder, "Millennials or Gen Z: Who's Doing the Most Job-Hopping," *CareerBuilder*, 2021, https://www.careerbuilder.com/advice/how-long-should-you-stay-in-a-job.

144. PWC, Millennials at Work Reshaping the workplace," https://www.pwc.com/co/es/publicaciones/assets/millennials-at-work.pdf.

145. Eric Schmidt and Jonathan Rosenberg, *How Google Works* (New York: Grand Central Publishing, 2014).

146. www.aauw.org/resources/research/the-stem-gap.

147. Megan Cerullo, "Nearly 3 Million U.S. Women Have Dropped Out of the Labor Force in the Past Year," CBS News, February 4, 2021, www.cbsnews.com/news/covid-crisis-3-million-women-labor-force.

148. Sarah Coury, Jess Huang, Ankur Kumar, Sara Prince, Alexis Krivkovich, and Lareina Yee, "Women in the Workplace 2020," McKinsey & Co. and LeanIn.Org, September 30, 2020, www.mckinsey.com/featured-insights/diversity-and-inclusion/women-in-the-workplace#.

149. McKinsey & Company, "Diversity Wins: How Inclusion Matters," May 19, 2020, www.mckinsey.com/featured-insights/diversity-and-inclusion/diversity-wins-how-inclusion-matters.

150. Sarah Coury, Jess Huang, Ankur Kumar, Sara Prince, Alexis Krivkovich, and Lareina Yee, "Women in the Workplace 2020," McKinsey, Sept. 30, 2020, www.mckinsey.com/featured-insights/diversity-and-inclusion/women-in-the-workplace#

151. Jane Kellogg Murray, "In 2020, 68% of U.S. Workers Say Their Job Became More Important Than Ever," Indeed, 2021, https://www.indeed.com/career-advice/finding-a-job/2021-job-seeker-survey.

152. McKinsey & Company, "Diversity Wins: How Inclusion Matters," May 19, 2020, www.mckinsey.com/featured-insights/diversity-and-inclusion/diversity-wins-how-inclusion-matters.

153. Diana Ellsworth, Ana Mendy, and Gavin Sullivan, "How the LGBTQ+ Community Fares in the Workplace," McKinsey, 2020, https://www.mckinsey.com/featured-insights/diversity-and-inclusion/how-the-lgbtq-plus-community-fares-in-the-workplace.

154. Glassdoor, "LGBTQ+ Employees Are Less Satisfied Than Colleagues at Work," 2021, https://www.glassdoor.com/blog/lgbtq-employee-satisfaction/.

155. Veronica Zambon, "What are some types of different gender identity?" *Medical News Today*, 2020, https://www.medicalnewstoday.com/articles/types-of-gender-identity#types-of-gender-identity.

156. Erica Hutchins Coe and Kana Enomoto, "Returning to Resilience: The Impact of COVID-19 on Mental Health and Substance Use," *McKinsey*, May 2019 report, https://www.mckinsey.com/industries/healthcare-systems-and-services/our-insights/returning-to-resilience-the-impact-of-covid-19-on-behavioral-health.

157. Jessi Hempel, "Elsa Sjunneson on Her DeafBlindness: Being disabled isn't a burden. The idea of not being welcome is the burden," LinkedIn *Pulse*, 2021, https://www.linkedin.com/pulse/elsa-sjunneson-her-deafblindness-being-disabled-isnt-burden-hempel/.

158. Ellen Francis, Helier Cheung and Miriam Berger, "How does the U.S. compare to other countries on paid parental leave? Americans get 0 weeks. Estonians get more than 80." *Washington Post*, 2021, https://www.washingtonpost.com/world/2021/11/11/global-paid-parental-leave-us/.

159. Government of Canada, "Employment Insurance maternity and parental benefits," Canada, 2021. https://www.canada.ca/en/employment-social-development/programs/ei/ei-list/reports/maternity-parental.html (accessed March 5, 2022).

160. Marc Benioff, Monica Langley, *Trailblazer: The Power of Business as the Greatest Platform for Change* (New York: Penguin Random House, 2019).

161. Rob Cross, Kevin Oakes, and Connor Cross, "Cultivating an Inclusive Culture Through Personal Networks," *MIT Sloan Review,* 2021, https://sloanreview.mit.edu/article/cultivating-an-inclusive-culture-through-personal-networks/.

162. Scott Osman and Marshall Glodsmith, *Leadership in a time of crisis: "The way forward in a changed world* (New York: Rosetta books, 2020).

163. Women's Business Enterprise National Council (WBENC), "Documentation Required for WBENC Certification Application," www.wbenc.org/documentation-required-for-wbenc-certification.

164. U.S. Women's Chamber of Commerce, www.uswcc.org.

165. Sarah Coury, Jess Huang, Ankur Kumar, Sara Prince, Alexis Krivkovich, and Lareina Yee, "Women in the Workplace 2020," *McKinsey*, 2020, www.mckinsey.com/featured-insights/diversity-and-inclusion/women-in-the-workplace#.

166. Eaton, Durme, and et al., Deloitte.com, Diving Deeper: Five Workforce Trends to Watch in 2021, *Governing Workforce Strategies,* accessed February 18, 2022, https://www2.deloitte.com/us/en/insights/focus/human-capital-trends/2021/workforce-trends-2020.html/#endnote-sup-17.

167. Luciana Paulise, "McDonald's Finds New Ways to Attract Young Talent through Amazon Alexa," *Forbes*, 2021, https://www.forbes.com/sites/lucianapaulise/2021/08/11/mcdonalds-finds-new-ways-to-attract-young-talent-through-amazon-alexa/.

168. Ryan Pendell, "7 Gallup Workplace Insights: What We Learned in 2020," *Gallup*, 2020, https://www.gallup.com/workplace/327518/gallup-workplace-insights-learned-2020.aspx.

169. https://www.gartner.com/smarterwithgartner/make-way-for-a-more-human-centric-employee-value-proposition.

170. Tony Hsieh, *Delivering Happiness* (New York, NY, Grand Central Publishing, 2013, 197).

171. David Pedulla, "Diversity and Inclusion Efforts That Really Work," *Harvard Business Review,* 2020, https://hbr.org/2020/05/diversity-and-inclusion-efforts-that-really-work.

172. Rob Cross, Kevin Oakes, and Connor Cross, "Cultivating an Inclusive Culture Through Personal Networks," *MIT Sloan Management Review,* June 8, 2021, https://cdo.mit.edu/blog/2021/06/09/cultivating-an-inclusive-culture-through-personal-networks-2/.

173. Luciana Paulise, "Amazon Offers to Pay College Tuition And Increases Minimum Wages To $18," *Forbes*, 2021, https://www.forbes.com/sites/lucianapaulise/2021/09/14/amazon-offers-to-pay-college-tuition-and-increases-minimum-wages-to-18/.

174. Luciana Paulise, "How to Lead a Remote Team With Vulnerability-Based Trust," *Forbes*, January 18, 2021, www.forbes.com/sites/lucianapaulise/2021/01/18/how-to-lead-a-remote-team-with-vulnerability-based-trust/?sh=63691573988.

175. Luciana Paulise, "6 Ways to Build a Culture in a Hypergrowth Company," *Forbes*, January 11, 2021, www.forbes.com/sites/lucianapaulise/2021/01/11/6-ways-to-build-culture-in-a-hypergrowth-company/?sh=7db9a3503143.

176. Carolina Valencia, "How to Get Employees to (Actually) Participate in Well-Being Programs," *Harvard Business Review*, 2021, https://hbr.org/2021/10/how-to-get-employees-to-actually-participate-in-well-being-programs

Part 4:
EMPOWERMENT

Control means giving people autonomy and agency over their work, empowering people with trust, and allowing them to make their own decisions- because they know their roles and responsibilities the best. Control can come in many forms: from choosing when to work (scheduling) and where to work(remotely or in the office) to creating your own job title (like the receptionist who was the director of first impressions) and deciding on the functions you fill.

—Jenn Lim[177]

T he fourth and final dimension in the CARE framework to drive a We Culture is empowerment: give team members the power and the tools to become more autonomous. Employees have to define new ways to get their jobs done no matter where they work, set goals by themselves, and agree on team routines to ensure team flow. The outcome is an agile organization that is able to react quickly to change while making employees feel proud of themselves at work.

Some people learn fast to work autonomously, while others may struggle. Some leaders find it difficult to empower. Particularly with remote workers, the tendency is to try to micromanage even more. But how do you micromanage remotely? It's simple—forget about it. Empower them instead.

What does empowerment mean exactly? Empowering means giving more autonomy to people in their areas of expertise.

Many leaders have a hard time empowering others because they believe that if they have to explain the task, it will take longer, or simply because they can do it better. The problem is that these statements limit leaders to do only what their time allows them to do. Empowerment, on the other hand, enables leaders to work through other people. Decisions can be made faster and at the right level, increasing ownership and accountability.

In the previous chapters, we worked on connecting everyone to a common purpose, paying attention to data and asking for more details, and respecting the diversity of ideas and backgrounds. The last step to thriving in a hybrid workplace is to embrace self-organization so everyone plays by the agreed-upon rules even when nobody is looking or controlling. Part 4 is about improving the process of time-management through self-management and empowering others to take action to achieve agility (see Figure P4.1).

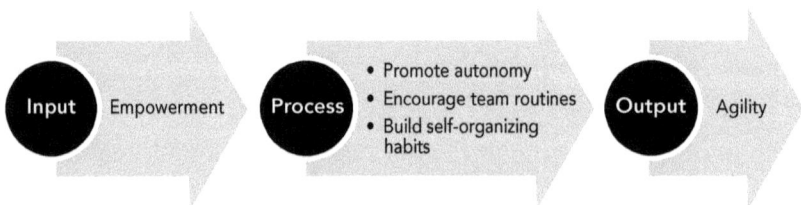

Figure P4.1   The process of time management in the empowerment dimension.

The three empowerment behaviors that will drive agility are:

- Promoting autonomy
- Encouraging team routines
- Coaching team members to build self-organizing habits

References

177. Jenn Lim, *Beyond Happiness* (New York: Grand Central Publishing, 2021), 53.

10

Skill #10: Promote Autonomy

If we all understand the "why" behind standards and truly own the processes, we don't have to be told what to do by our leaders.

–Tracey Richardson,
The Toyota Engagement Equation[178]

D elegation is a skill usually taught in leadership courses, which usually just means telling someone to do part of your tasks that you don't like.

The focus here is not on teaching the leader how to delegate but on how not to be the owner of the task in the first place, and help gain accountability. The idea is to foster individual responsibility while leaders give away control.

The people closest to the production have to be empowered to make choices about how they are going to spend their time based on their knowledge of the work, its difficulty, and their individual abilities. There has to be a system in place that makes everyone on the team trust that they can choose the solution they think is best and that the leader or other team members will support them.

Key Insight

The people closest to the production have to be empowered to make choices about how they are going to spend their time.

Usually, empowering is confused with delegation. We want to teach leaders to delegate better, and they tend to go to the extremes of the delegation spectrum: Many leaders understand they just need to "get rid of tasks" and put the burden onto someone else while giving little advice (because they need to "get tough"), while those on the other extreme only allow their team members to do the more operational part of the work while they keep the fun part and, of course, monitor every movement. These are the abandoning leaders and the micromanagers, respectively. Empowering is much more than that; it has to do with giving the team members a complete task to do that they can own, make decisions about, and feel responsible for.

Empowerment enables leaders to work through other people.

Me Culture Behavior: Micromanaging

Micromanagers are leaders who want the job done their way but provide little advice. Leaders tend to micromanage more than usual during a hard time, but that doesn't mean they help more.

Micromanaging looks like constant individual check-ins, long team meetings, or multiple daily meetings where only the manager speaks. They are continually asking for the status of tasks on random occasions. They only provide feedback after the job is done, focusing primarily on the weaknesses and improvement opportunities. The when and how-to of an assignment is not agreed upon as a team. Every decision is made top-down and has to be executed at the pace defined by the leader. In the long term, there is no relationship of trust because the leader feels there is no need for trust. Employees just need to get things done as instructed.

Employees in turn feel frustrated and are usually disengaged at work, as they don't find ways to collaborate or speak their minds. They work a lot to try to satisfy the changing demands of the leader, but they never have a clear picture of how the work should get done. Problems and tasks, in the end, seem to belong to the manager, not the employee. And the worst part is that creativity and improvement are reduced, as team members don't feel valued. They tend to stay quiet and follow instructions. They either have low performance or leave the team.

Me Culture Behavior: Abandoning

On the other end of the spectrum, there are leaders who trust their employees too much that they don't even bother to check in. They only have the required monthly meetings to report results.

Employees are better able to manage their time and how they get the job done, but again, employees feel frustrated at the end of the day. When they get to talk to the leader or receive feedback, the job is already done, and there is nothing they can do to improve what happened; they can only improve in the next assignment. Thus, feedback sessions hardly ever felt rewarding.

This leader may seem trustworthy, but the problem is that technical aspects of the job may not require continuous feedback from the leader. The personal side of the employee has been pushed aside. And, in case technical feedback is offered, it's delivered too late. Just like with micromanagers, they fail to provide timely advice.

Abandoning leaders are very common, especially in hybrid workplaces, where leaders tend to focus on the employees they can see and abandon the ones they don't see often. They simply say, "Contact me any time you need assistance," but they don't plan on having periodic meetings. Some employees may ask for help, but others don't in the name of fear or pride and internally feel set aside and less worthy.

Problems seem to belong to these employees solely, so they must solve them on their own. Team members have frequent thoughts about leaving the company or fear they may not be needed anymore. If they are terrified of being laid off, they may work harder for some time trying to compensate, but this won't last long. In the long term, these employees get the job done but avoid presenting new ideas or challenging the status quo, as they don't feel rewarded.

We Culture Behavior: Facilitating

The role of the leader is not to micromanage or delegate all the tasks; rather, it's to facilitate interactions. If the leader is slowing things down for the team, something is not right.

Key Insight

The role of the leader is not to micromanage or delegate all the tasks; rather, it is to facilitate interactions.

The leader's main tasks are:

- Improving communications
- Eliminating other meetings or distractions
- Identifying impediments
- Highlighting and promoting quick decision-making
- Improving the team's level of knowledge

Foster Autonomy

While management provides the main rules defined in Skill #1, the teams need autonomy to follow those rules. Autonomy is crucial, but it is also necessary to provide support and celebrate progress. Instead of keeping ownership of the tasks, or giving up on them completely, leaders have to let the team own those tasks. Leaders' focus must be on promoting two-way communication with team members, within the team, across the company, and with customers, suppliers, and the community.

Leaders need to help team members start the change and create a face-to-face or online environment where people feel safe providing updates, offering ideas, and asking for support. Make it more of a "pull" system where they ask for help, not a "push" where they are told what to do and when.

Key Insight

Make it more of a pull system where they ask for help, not a push where you tell them what to do and when.

Encourage employees to set up goals and take ownership of their work, moving yourself to a position where your own goal is to unlock their potential and embrace technology to find new ways to set up your teams for success.

As we have seen in Skill #4, making decisions with uncertainty is like planning long term: everything is variable. However, when you empower your team, they will become sensors to help detect and deal with issues in the future more quickly than if you had to do it all yourself.

Key Insight

While you empower your team, they will become sensors to help detect and deal with issues in the future more quickly than if you had to do it all yourself.

Empowerment enables leaders to work through other people. Strengthening employees' autonomy and empowerment makes it easier for employees to work on their terms and for leaders not to babysit or micromanage.

Autonomy is key to ensuring employees respect the company's rules, even when "nobody is looking"; that's the ROI of a strong culture. 1–800 Contact promotes empowerment by avoiding scripts in the call center and allowing employees to decide how many hours per week they will work. This policy allows the company to hire employees from remote locations, even students or people with disabilities. Phil Bienert says autonomy is a blessing when employees need to work remotely, as they learn to make more decisions by themselves. 1–800 Contact also allows employees to decide how to WOW their customers, even if it entails paying for an Uber ride or sending unexpected gifts.

Watch the interview with Phil Bienert
https://youtu.be/rpXVAPNZjVg
or download the We Culture app to learn more.

Autonomy increases engagement and job satisfaction, as employees feel more unique and valued. The other benefit of autonomy is that it makes employees less dependent on the leader and more self-organized, reducing both team member and leader stress and frustration while working from home.

Set Clear Expectations

The values and goals of the organization need to be connected to the values and goals of the employees. The role of the leader is to help the employee make this connection by discussing and setting the right expectations. Workplaces that give their employees greater freedom to complete their work experience greater job satisfaction. Millennials in particular are more productive if they have more autonomy.[179]

1. Share the context and goal of the team.

2. Define individual roles and responsibilities.

3. Agree on goals and deadlines.

4. Offer help.

Share the Context and Goal of the Team

This is the big picture, and it is essential for employees to know how their work fits into the team's goals. What are you trying to achieve? Why is it important? How will it affect customers?

See Skill #1, Define a shared purpose and core values.

Define Individual Roles and Responsibilities

Even when defining roles and responsibilities seems to be a well-known issue, the Anatomy of Work index found that 29% of knowledge workers feel overworked due to a lack of clarity on tasks and roles, and a disconnection from the goals of the organization. Define the tasks that will be done by the employee, or show the total number of tasks in a team meeting and ask employees to decide who will do what. Define owners for each task to accomplish the team's goals, and make sure to keep these responsibilities on a board or online tool so anyone can check it at any time.

There are two main rules:

- Every task has an owner.
- No task has two owners.

See Skill #1, Define a shared purpose and core values.

Goals and Deadlines

Define clear goals, deadlines, milestones, and updates instead of regularly checking in on your employees unannounced. Start trusting that your employees can do their work successfully while working remotely without constant oversight. Discuss how employees will know if they are successful.

Also, help them prioritize these tasks. To avoid feeling burnout or lack of accomplishment, every individual should plan one to three tasks to accomplish each day; focus on getting them done by the end of the day; and prioritize these over last-minute demands or small tasks like email and chat. When completed, this should feel like permission to "turn off" and stop working for the day. See Skill #5, Set your own quality goals.

Offer Help

Address concerns by asking questions. Make sure employees know whom to contact or when if they need help.

The details to provide regarding how to get the work done depend on the skills of the employee and the type of job. With a very skilled employee in a self-organizing team, leaders are facilitators who simply coordinate a meeting to distribute tasks and provide the context, the customer needs, and the project goals (step 1). The other steps are defined as a team.

In a group that is still very leader-dependent, the four steps are determined by the leader. To become more autonomous, leaders need to find ways to move these steps to the team.

Leaders need to help team members improve self-organization and build the right context that enables empowerment to empower team members. Again, it is not just about giving a task but also about giving part of your power to own something.

Leaders empower their employees by giving them the tools and support to perform their jobs.

See Skills # 7, 8, and 9.

As you can see, the skills we have been reviewing help develop an environment where fostering autonomy is easier: the We Culture.

Build Trust

One of the main reasons why companies prefer working 100% in an office is that they don't trust that their employees can work from home effectively. If this is the case, the problem is really not about whether the employees are performing but whether the leader will trust them.

You need to know and let your team know that you are not perfect, and neither are they. A continuously changing environment requires people to trust and learn from others. Of course, that doesn't make us weak. It does, however, require that we create a safe space to make mistakes and trust that everyone is doing their best, even when they fail, and even when we fail.

Trust yourself and get past "impostor syndrome," feeling that you are not good enough. If you don't trust yourself, you will hardly trust others. You are there for a reason, so simply do your best. It doesn't have to be perfect, your OKRs don't have to go all green. You will do even better if you ask for help from your team.

Employees Can Be Trusted to Do the Right Thing

Different industries view numbers and mistakes differently. Failure seems to be normal in the IT industry, where companies like Google, Apple, and Facebook were born with a new mindset. Errors are OK as long as they bring innovation in the long term. But this doesn't seem so accurate in a manufacturing or accounting firm. Still, employees should be trusted.

The European company FAVI is a manufacturing company[180] that got rid of production control systems and hourly targets. Not surprisingly, when it changed its system, productivity increased. Workers started to work in a level that was more natural for them and therefore more effective. Employees knew better. In many manufacturing companies I worked for, employees were not even trusted to get the stocks they needed without approval. This reduced their productivity, as in many cases they needed to wait for their supervisor to change a cutting disk.

The more you emphasize numbers, the more the operators will try to cover up problems and try to take advantage, just in case.

Key Insight

The more you put emphasis on numbers, the more the operators will try to cover up problems and try to take advantage, just in case.

Imagine how much you can save when you redistribute effort because you don't need to control employees. How much are you spending on controls? Measure how much you trust your employees by tracking how often you tell people what to do versus asking them for input.

Reverse Delegation

FAVI implements what is called "reverse delegation." The book *Reinventing Organizations* describes it as "the expectation is that the front-line teams do everything, except for the things they choose to push upward," meaning they create a meeting, a role, or a process only when the opportunity arises. A new coordination function may come from a team decision of reverse delegation. However, most of the coordination happens on the shop floor.

Is this possible in hybrid or remote workplaces? Yes. At Buurtzorg, a nursing company with more than 7,000 nurses scattered across the world, most of them have never met, and they work this way.

Use Technology to Help Team Members Become More Autonomous

Teams can take advantage of technology to help them organize tasks, track performance, and share project status. As mentioned in Skill #5, one of my favorite apps to build routines and promote autonomy is Sunsama. This app was developed to enable individuals to keep daily and weekly routines to organize, prioritize, and execute tasks in the most productive way.

The app prompts a routine every time you turn it on, every day. It starts by asking what you want to get done today. It has the ability to bring tasks from various sources, such as apps, calendars, or emails, which at the end of the day make it complex to combine. Sunsama compiles these apps for you on one screen.

Then the app asks what tasks can wait and be bumped to the next day to avoid burnout or create stress. To avoid an unpredictable workload, it has the option to add a "planned duration" for each task, and it warns you when you have too many tasks without an estimate.

Finally, it prompts you to arrange the tasks in the order you want to work on them and set a shutdown time.

Once you are ready to start working, you should start with the first task on your list and push "play" to measure how long you are working on it. It will also help you focus on that task by reminding you to work only on that task.

When it's shutdown time, it will remind you to wrap up your day, tell you how long you worked and what tasks you accomplished, and ask you to document why it was a good day.

This routine helps you plan so you stay focused on one task at a time, flow until you finish your tasks, and reflect at the end of each day.

Once you define your own routines, help your coworkers define team routines to coordinate work more easily. Use the same apps if possible to be able to check on each other when needed and speak "the same app language."

Manage Time the Agile Way

One of the most difficult parts of empowering employees to achieve agility and self-organization is making sure things get done right and on time. When employees get used to following orders or being micromanaged, they don't learn to manage their time. Some are perfectionists by nature, spending hours on the same task, while others simply dedicate as much time as the leader pushes. But being busy isn't the same as being effective. Team members may work a lot but not finish the tasks or spend more time on the wrong activities. Everyone on the team should learn how to prioritize and define how long to spend on each activity, and consistently apply a system to allocate tasks into their daily schedule. Time management refers to the way you create this routine

Following a set of behaviors, the 5Fs, will help you manage your time and the time of your team more effectively (see Figure C10.1).

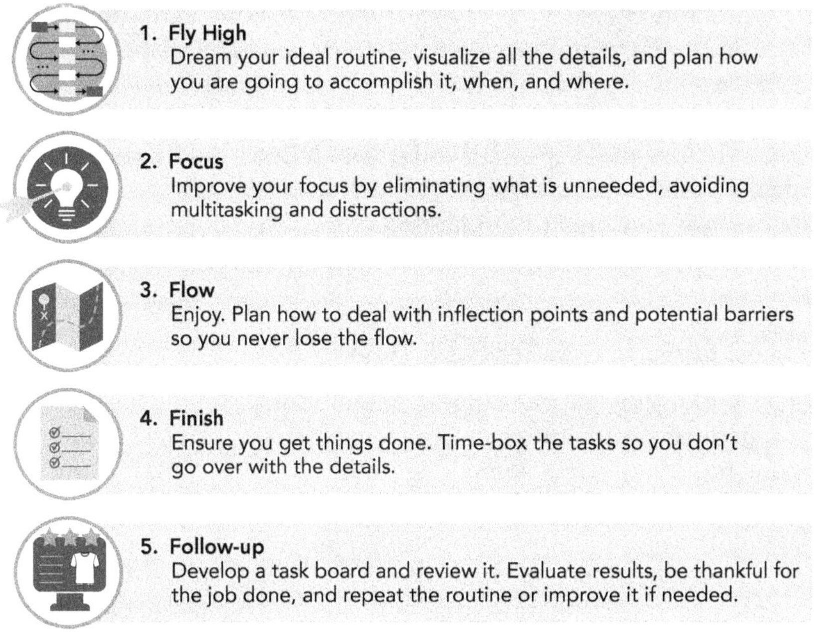

1. **Fly High**
   Dream your ideal routine, visualize all the details, and plan how you are going to accomplish it, when, and where.

2. **Focus**
   Improve your focus by eliminating what is unneeded, avoiding multitasking and distractions.

3. **Flow**
   Enjoy. Plan how to deal with inflection points and potential barriers so you never lose the flow.

4. **Finish**
   Ensure you get things done. Time-box the tasks so you don't go over with the details.

5. **Follow-up**
   Develop a task board and review it. Evaluate results, be thankful for the job done, and repeat the routine or improve it if needed.

Figure C10.1 The 5Fs of time management.

Fly High: Plan

Visualize and plan all the tasks that need to be done during a period of time. You have to cultivate the habit of imagining, as precisely as possible, what you expect to see and what problems may arise, so you consider as many tasks as possible from the very beginning. Start with next week. What do you need to get done? Define how many hours you want to work a day and how you want to go about breaks, snacks, and free time.

Define no more than three main goals for the week. Now, divide those projects or goals into smaller parts, that is, tasks no longer than one to three hours. That way, you don't get lost in a big project, and you are able to make adjustments more quickly.

Now plan your first day of the week: what are the tasks you will accomplish on Monday? Apply the 80/20 rule: 20% of your activities will account for 80% of your results. Therefore, if you have a list of 10 items, by doing just two you should be accomplishing 80% of your results. So, what are those two?

Add the events you have in your calendar and combine all your sources of to-do lists so you have a combined list of things to do. Sometimes we plan to do a number of tasks while we forget to contemplate that our day is full of meetings. Prioritize the tasks in the to-do lists and respect the order you defined. Add time estimates to your tasks to get a better sense of your workload. Only add new items once you finish the planned ones. If there are urgent issues to attend to, prioritization should help you define what matters the most.

Focus: Perform One Task at a Time

Start working one task at a time. Don't multitask; follow the priorities you defined before. Stephen Covey used to say, "Unless something more important—not something more urgent—comes along, we must discipline ourselves to do as we planned." Just focus on the task at hand.

Eliminate what you don't need to accomplish the task. Avoid distractions by eliminating from your sight all the tasks, paperwork, noises, calls and messages, social media, food, and especially clutter, and focus all in. Don't worry about what comes next.

Don't say yes to every external demand. Focus on following your own plan first. If you usually have many external demands from customers or employees, set an unscheduled time in your calendar to deal with those, but don't do it while you are working on one specific task.

Flow: Enjoy Each Task As If It Were the Only One to Get Done

Flow is a special state of mind, which moves people to do their best work, no matter what work they do. Have you ever lost track of time while doing something you loved? That is the state of flow. The challenge absorbs you so much that you lose track of time. This step's end goal is to flow in your designed routine and follow it without thinking. Tony Robbins sticks to a morning routine designed to boost his energy and productivity levels for the day. Mark Zuckerberg's "work uniform" consists of jeans, sneakers, and a gray T-shirt, so he doesn't have to stop and think about it. That's the key! Don't even think about changing the most important tasks until you get to achieve them.

Set a timer to start your flow. Every time you start an activity, be mindful of that activity, and stay focused by starting a timer and stopping the timer once you finish. That way, you don't just focus on doing that task, but you also will be able to measure how much time you spent doing it. When we are in flow, we may forget about time, but that could be counterproductive when having many other things to do. The secret of time management is limiting how much time you dedicate to each activity, even breaks or snack times. Good sprinters are one thing above all: in flow. They simply run as fast as they can to win the 100 meters; they don't even look at the watch, they just run. Don't even evaluate how you are doing because that's a distraction in itself. Keep moving until time's up, and then you can evaluate your results and adjust for next time.

Plan how to overcome barriers: If you have a remote team, you want to do a daily meeting at the same time every day. Some unexpected issues may arise: you get caught on a phone call or your main report is not ready for the meeting. Plan in advance how to deal with these inflection points to limit stress and continue with the routine. Then just enjoy: flow is a pleasure. Repeat, repeat, and repeat the routines that help you reach the state of flow, until you do them naturally without thinking, and they become a quick win for you and your team.

Finish: Get the Task Done Completely Before Going to the Next One

The whole point of getting things done is knowing what to finish and what to leave undone. Sometimes it is hard to finish tasks because you get distracted or second guess yourself. You start emails that you never send or start an analysis that you never finish. What's wrong? There are three ways to avoid this in the future: 1) time-box or allot a specific time for each activity; 2) book it on your calendar so you don't go over perfectionism; and 3) accept what you have done; perfection is the enemy of action. Author James Clear says, "If I have to write an article every Tuesday, it doesn't matter how good or how bad I feel about the article; I try to do my best but stick to the schedule anyway. Sometimes it is just that you are too hard on yourself."

Get to done: Make sure you check it off your list. Avoid "almost finished" activities like an email written but not sent. They fill your calendar with no impact on results. This happens to team members too, especially when leaders are micromanaging. Make sure you and your leaders are not micromanaging employees and allow them to finish their own tasks.

Follow Up: Get Better Next Time

Evaluate results at the end of the week and take action to improve your priorities and task duration next time. If your schedule worked well, repeat the same routine: consistent practice produces mastery and makes for new habits. Which tasks took longer than expected? Which tasks were unplanned? Which tasks could be reduced? Which tasks could be delegated? Were there any extra costs? By the end of the day, be thankful for all the tasks accomplished. Don't get frustrated focusing on what went wrong.

Failing to manage your time is not like a genetic trait, something you cannot change. It is the result of strategy and discipline. The self-discipline to keep a routine is like a muscle; it has to be trained. Start today with these 5Fs and accomplish more with less!

Hands-on 10.1

Follow the 5Fs to plan your pending tasks and identify what you can maintain and what you can do to improve.

Individual Exercise

Follow the 5Fs to plan the tasks you have pending, and identify what you can maintain and what you can do to improve.

To maintain	To improve
_____	_____
_____	_____
_____	_____

References

178. Tracey and Ernie Richardson, *The Toyota Engagement Equation*, Indian edition (India: McGraw Hill Education, 2018).

179. PWC, "Millennials at Work Reshaping the Workplace," https://www.pwc.com/co/es/publicaciones/assets/millennials-at-work.pdf.

180. Frederic Laloux, *Reinventing Organizations* (Brussels, Belgium: Nelson Parker, 2014).

11

Skill #11:
Encourage Team Routines

Your daily planning ritual is the force multiplier you'll use every single day to make sure you work on what's important. This is your chance each day to decide what work moves you forward and what is just bullshit.

–Ashutosh, Sunsama founder[181]

C ompanies can get a customer-service-driven culture, engaged employees, and a great brand by building the right routines.

A team routine is a sequence of actions the team follows periodically to accomplish a result, such as a daily meeting at the end of the day. The team utilizes routines to make the interactions among team members easier. Meetings, end-of-day reviews, or workplace cleanups are examples of team routines.

Team members need to ensure there are systems in place for every activity that needs to be done, so everyone executes the activity applying the best practices. Some systems require very detailed standards; others simply need to guide the employee. Their absence confuses employees and makes them more dependent on leaders. Leaders should be facilitators, not problem solvers. For example, one common "disease" nowadays is the immense amount of time spent during meetings. Team meetings are not bad—they actually save a lot of time if they are carefully planned. When teams don't have pre-planned meetings, however, they constantly contact individual team members to ask for help or solve problems, disturbing the employee workflow and preventing teamwork. Team meetings instead help to prevent interruptions by making sure everyone has a time set aside for questions and discussions. The problem is when team members or leaders spend most of their time attending meetings and not getting things done.

Agile teams prefer fewer meetings that are more results-oriented, and probably shorter and more frequent. Lots of companies like Toyota have daily stand-up meetings instead of weekly meetings. A routine could be having one stand-up 15-minute session to communicate quick updates across shifts when every shift changes. Now, suppose the routine is not carefully defined, and the 15 minutes becomes one hour. In that case, invitations are random, people become tired of spending so much time so they skip the meetings, and others don't attend prepared. Then the routine is no longer useful.

Me Culture: Relying Too Much on Complex Standards

Standards and routines should be designed to help, not to interrupt, the flow. Teams usually see systems and standards as part of a "bureaucracy," such as the aforementioned meetings or big company manuals, so wordy that no employee ever reads or understands them. Bureaucracy is a way of safeguarding from mistakes. In many cases, it ends up being impractical and costly.

Systems should help employees recollect the most effective ways to act or learn a curated routine, not only for themselves but also for the organization. Routines must be easy and available for employees to follow. Standards help people follow routines, and routines help teams be more effective. Standards can be lengthy procedures or simple pictures, arrows, and colors.

Most companies nowadays already have procedures for their most important tasks. But for some reason, when you observe employees doing their jobs, most of the time, each employee does it differently. Be encouraged to ask employees where to find the specific details of the task, and often each employee responds with a different answer. Why does this happen? What about the standard? They are used to following instructions by observing others, not by reading standards or manuals.

When standards are not met, or there are no standards, processes usually create wastes. But when there are too many standards that are not used, there is a lot of waste too.

Besides, the rules tend to be slightly modified over time. It's like when you try to remember a movie some of the details are blurred in your mind. When you do the same activity over time, you adapt the procedure for your convenience, almost by accident. Changing the procedure for convenience can be a great idea, but why not make it official?

It's just that procedures aren't usually stored near where the work is done. They are probably in a large folder inside the office, on a shared disk, or in offsite storage. And if they are available, updating them is usually not an "employee-friendly" practice. Maybe a consultant does it three years later when it's time to update the ISO.

Generally, employees are not empowered to review or update the rules as a matter of daily practice. Even so, processes are always changed by employees on an informal basis. It doesn't make sense, does it? Actually, it makes perfect sense. Employees, over time, understand the tips and tricks of their process. They know better how to do it.

Have you seen babies eat? Through practice, they learn how to eat, use their hands more effectively to get more food in less time, order food, and throw food away to complain. It's part of their development.

The practice is part of the employees' development until they become experts. Hence, it makes sense that they should be empowered to communicate their improvements in a constructive manner and take responsibility for their own work,

What is the point of these manuals if you cannot read or update them at any given time and should they not be available, especially to new hires?

Employees know what they need. They should be responsible for developing the standards and keeping them up to date, relevant, and posted at the workstation or available for anyone who may need them.

It's all about convenience.

Make it convenient to find, easy to follow, and easy to update.

The procedures and rules do not have to be in a heavy and complex manual to be good.

You can save the massive manual and then simply have pocket editions or more visual versions on the workstations. Standards may look like pictures, checklists, manuals, schedules, drawings, colors, or lines on the floor — anything that helps clarify what needs to be done.

It is necessary to facilitate compliance by making the standards visible to everyone on the workstation and available for easy retrieval.

When something doesn't appear to be right, team members need to ask the following questions:

- Is there a standard for the situation?
- Is there a standard, but it is not being followed?
- Is the standard correct?

Problems with the routines should be put forward right away to avoid them being hidden or postponed. Hidden problems simply create more work in the future. As long as associates find problems and look for help to solve them, they will be working on avoiding similar issues in the future.

When we don't have routines or we don't use them correctly, we need to make more decisions, and more decisions means more variability. And variability is the opposite of quality: it means more errors. The purpose of quality is to reduce variability so customers always receive what is promised.

We Culture: Setting Team Routines

The same way companies have procedures and systems for manufacturing a product or performing a service, they should have defined systems too, for instance, to make meetings more effective, to clean up the workplace, or to provide feedback. If standardized across the company, these team routines will help instill the new behaviors this book has been describing, such as active listening and equality. In addition, these daily interactions account for most of the activities we perform during the day, so why not be intentional about them?

While the customer experience may depend on how the service or manufacturing processes work, the employee experience depends on how team routines work.

Team routines help teams do collaborative work more efficiently. For example, you can establish how to:

- Organize meetings
- Communicate
- Manage projects
- Solve problems
- Make decisions

They provide just enough structure to allow a team to run without external coordination.

If predefined and agreed upon by all team members, team routines help establish the right amount of synchronous and asynchronous communication. Especially if working from home, you need to define specific times or methods to communicate to help organize individual and team priorities. For example, if the goal of a meeting is just to share information, consider defining a routine for emailing these types of presentations. On the other hand, feedback sessions are better as individual, face-to-face meetings that also should be planned and orchestrated to drive the best possible results, every single time (see Skill #7). Again, the idea is to make it more simple and frictionless for the employees to do their job.

The good thing about making routines a habit is that they help us avoid making conscious decisions all the time because we don't ask ourselves whether it is OK, we just do it. They help us be more effective and focus only on what needs our attention. Instead of worrying about how to set up a meeting or how to solve a conflict, just focus on finding the facts. For the rest, there is a routine in place that you just need to follow.

The bad thing about habits is that if those habits are not helpful or smart, they become part of a vicious cycle driving negative results, which can be a big burden in our lives (like eating too much, drinking too much, or spending too much).

A great example is swimmer Michel Phelps, who won 28 Olympic medals thanks to the power of a routine. When he was swimming as a child, he had problems coping with stress. To manage his stress, he used to have the same routine every day before the competition to help him calm down. Every morning and every night, he would "watch a videotape," like a visualization of every detail of the race, until he knew the tape by heart. He visualized how he was going to eat, get dressed, put his goggles on, and get ready. He also planned how he would jump and how many strokes he had to take to get to the other side. He would plan every detail of his routine and what could go wrong and prepare for it every day the same way. An intentional daily routine made him unbeatable.

One day when he was swimming for the gold medal in Beijing, his goggles began to leak. He couldn't see where he was swimming, but he just followed the videotape in his head, his routine. The result? He finished with a world record anyway.

Key Insight

An intentional routine can make you unbeatable over time.

The 30-Minute Meeting Challenge

Meetings could be the most productive part of your day, as well as the least productive. The C-suite typically spends approximately 72% of the time in meetings,[182] most of which last at least an hour or more.

That is why it is crucial for teams to have a very specific routine for team meetings, either online or face-to-face, to ensure all the topics are discussed timely and everybody attending can participate. If someone doesn't participate, he or she should not be in the meeting or the meeting is not facilitated correctly. A routine for a hybrid team can help, for instance, to ensure both onsite and remote workers equally participate.

When a meeting is well organized, you can get lots of ideas, energy, and action plans out of it. Especially if you are working remotely, meetings are the perfect time to foster the personal connection, share issues on a timely basis, and explore suggestions.

But you can also get employees' time and attention stuck with no result. Hundreds of unproductive hours a year per employee are lost waiting for other attendees, looking for documents during a meeting, repeating topics, extending discussions without agreeing on a clear action plan, missing key decision-makers, or holding attendees who don't participate. And if you also consider the time spent before and after the meetings, your unproductive hours can be exponentially high.

As a leader, team member, or facilitator, it is your role to ensure meetings are the most productive part of the day for everyone involved. Your team will develop better ideas, increase engagement, and improve collaboration, while you will be reducing your meeting time significantly. How a leader manages time demonstrates their leadership style, what that leader prioritizes, and how he or she communicates. Assessing the quality of meetings, prioritizing who should attend, and ensuring attendees' complete focus will drive success in the short term.

Prepare the Meeting Schedule in Advance

For periodic meetings, always follow the same agenda order. For example:

1. What did I do yesterday?

2. What obstacles stand in my way, do I need help?

3. What will I do today?

Everyone should get ready to answer these three questions or whatever basic agenda you define. If the number of attendees is also fixed, it should be easy to estimate how much time everyone speaks.

Prioritize Who Should Attend

An HBR research[183] shows that engagement typically decreases the more time people spend in very large group settings. People would be more productive and engaged if they could spend less time in meetings and more time preparing for them.

Minimize the Number of People

Reduce the number of attendees to the Jeff Bezos two-pizza rule: no more than six to eight people. The more people, the more unproductive the meeting becomes. It also avoids wasting the time of those who shouldn't be there. Only people who can contribute should be invited, and they should be prompted to provide candid feedback. That's the secret of Pixar's Brain Trust meetings[184] to foster quality and innovation. Some self-organizing companies make the meetings public. They can be attended by whoever is deemed necessary. If the session is informative, it can be replaced by a video-on-demand, or an online tool can be used to collect data or questions such as Kahoo.it or menti.com.

Encourage employee confidence, self-discipline, and psychological safety (see Skill #8). Give team members more opportunities to choose which meeting to attend and how to participate actively when they decide to be part of a session.

Review your weekly meetings and ask yourself: Is this meeting necessary? What is the objective? What is your role? Do you need to attend or can you review the minutes? Especially working remotely, you may want to dedicate more time to one-on-one meetings than large ones. They are more meaningful, as they increase employee engagement, sense of belonging, and human connection.

Increase Focus

When meetings are too large or too long, and agendas are not concise, attendees tend to switch to offline mode, and turn the focus to emails or messages. That's the worst kind of wasted time. You are not able to focus entirely on emails or the meeting, and the other invitees don't get your input. A study conducted by Microsoft[185] shows that after the 30-minute meeting, the brain experiences excess fatigue, making it very difficult to concentrate.

In a webinar, it is even worse; people can pay the most attention for 10 minutes if there is no interaction. After that, either make sure there is a question or an engaging interaction, or you will lose your impact, especially for virtual meetings. Make it short and sweet to keep the focus on the issues at hand. Take advantage of online tools like Mural.co or Asana to increase engagement during the meeting, or to communicate results, request feedback, or assign responsibilities after the meeting.

Dedicate a specific time of the day to hold meetings and have them all in a row to reduce time fragmentation.[186] Time fragmentation is the time you lose when you multitask. It takes at least 15 minutes to become productive again after a break, and it takes at least 30 minutes to focus on a particular issue to move forward or make a decision. Some companies have free-meeting days, such as a Wednesday, when nobody can schedule meetings. Others have "no meetings after 2 p.m." policies.

Invite Collaboration

Create guidelines that facilitate discussions in Zoom so no one person takes too much airtime. You can also train people to always go on mute when not speaking.

Before closing the meeting, add the routine to go one by one, asking all the participants if they have anything to add. In hybrid environments, this helps ensure nobody is left out of decision-making. Even though everyone can virtually participate, the shy ones may not do so unless they are asked. Make sure you do.

Use technology tools like Trello or Asana to help assign tasks to employees clearly and make sure there is only one owner per task.

Clean-up for Respect

In face-to-face meetings, include a picture of how the room should look (or a checklist) so that before everybody leaves, the space remains organized and clean. Then, as a leader, set the example and be the first to clean up your spot.

Time-box the Agenda

Agendas and minutes are essential, but they are not effectively implemented nowadays. If you think about the last boring meeting you had, you will probably realize that topics were added at the last minute to the agenda. Some members spoke too much, while others were not even able to talk. It is most likely that a difficult topic took longer than expected because speakers were unprepared or information was not handy. That meeting could have been held in 30 minutes tops.

Create guidelines to facilitate discussions so no one person takes too much airtime. You can also train people to always go on mute when not speaking.

You can organize a short meeting every morning at 8 a.m., for example, so you get everybody aligned at least once a day. Making everyone avoid using phones and computers in a face-to-face meeting or turning on the cameras in a virtual one are excellent ways to ensure everyone is focused and not multitasking.

Make it a company priority to set a new meeting standard of no more than 30 minutes that will help your teams prioritize topics, select attendees more carefully, and become better prepared for the next meeting. You will be getting back at least 50% of your precious time.

The following are some other rules you can add to your meeting routine:

- Hold team meetings once a day, no matter what.
- Allow no more than 15 minutes for stand-up meetings.
- Send agendas two days before the meeting or they will be predefined (such as daily Scrums).
- Don't allow side conversations.
- Agree to share concerns openly.
- Use the consensus process for all key team decisions.
- Use company purpose and values as a decision filter.
- Assign owners to action items and write a timeline on the whiteboard to review the next day.
- Leave a chair empty for the "voice of the customer."
- Record all meetings on Zoom and store them in a specific meeting/project folder.
- Do not use a physical board. Share everything with online tools for remote workers.
- Add one minute, in the end, to relax and breathe before going to the next meeting to keep the attention at the highest level.

Whatever rules you decide to add to your routine, follow them no matter what. A routine only sticks to the employees' behaviors when practiced every single time until it becomes unconscious.

Applying SMED to Improve the Productivity of Your Team Routine

If you can't seem to reduce the length of your meetings, use single-minute exchange of dies (SMED). Formula 1 cars use it to improve tires' changeover time. For many people, changing a single tire can easily take 15 minutes. For a Formula 1 pit crew, changing four tires takes less than three seconds.

The proposed solution is to SMED your meeting. SMED is a method widely used in manufacturing to improve changeover time. In Ford-T times, all cars were black to avoid spending two weeks in changeover time to use a different die. Until Toyota's engineer, Shigeo Shingo, came up with SMED to reduce the process to three minutes.

Experience shows that changeover time can be dramatically reduced by as much as 94%.

Airlines make money when their planes are in the air. They do not make money waiting at a gate. Southwest can "change over" an airplane between flights in less than 30 minutes. You don't make money when you are waiting for a meeting to start either, so follow these five steps:

1. **Identify a pilot meeting.** Choose a periodic meeting that you consider inefficient and focus on improving it first; don't start with all the sessions at once. Choose a team that will help improve it. It is recommended that you have some of the current participants and some external participants or facilitators to help you figure out what could be done differently.

2. **Identify elements.** Have the team dissect all the elements of the meeting: discussions, people talking, voting time, note-taking, waiting time, silence time, idea generation, etc. The most effective way to do this is to record one of the meetings with your phone, have observers take notes, and then work from the recording to create an ordered list of elements. The elements should include the description of what's being done and the cost in time spent. Observers can also identify attendees who are not participating, are distracted, or are multitasking, and estimate time lost waiting for the meeting to start or employees who are taking over most of the time or intimidating others (hippo effect), for example.

3. **Identify external tasks.** Identify those tasks that could be done at another time, not during the meeting. External tasks could be done before the meeting, such as retrieving documents, approving previous meeting minutes, getting information, contacting employees who are not present, or preparing the agenda with the topics to discuss. Some

tasks that could be done after the meeting are discussing issues in detail, preparing the minutes, or checking the status of a report. Most of the time lost during sessions is part of these external elements. While they are in progress, attendees are waiting, and time is lost for everyone, while it could only be spent by one of the team members or eliminated.

4. **Convert internal elements to external.** Analyze the meeting to convert as many internal elements to external elements as possible. A meeting that includes several team members, including managers, should be reduced to the minimum number of internal tasks, while the rest of the time should be spent individually in preparing and reporting the external tasks.

   Decision-making or idea generation are usually internal tasks that need to be done during the meeting, but how can you prepare steps in advance to do them faster? For example, you can define or standardize simple voting methods, routines, or decision-making processes in advance.

   Many companies do stand-up meetings or daily 15-minute meetings. The routine is always the same: every employee reports, at the end or beginning of the shift, what is done, what needs to be done today, and the potential issues. This routine teaches attendees to be prepared and precise, sharing their updates in less than a minute, with a short-and-sweet speech.

5. **Streamline remaining elements.** Have the team ask the following questions for each element: How can this element be completed in less time? How can we simplify this element? For instance, time may be lost in remote teams due to system issues. You could prevent it by ensuring you have the right tools when needed. You could provide information and links in advance so attendees can download the tool, test it, and be ready to use it during the meeting. Zoom, for instance, has incorporated various apps, such as Mentimeter or Mural, used for collaboration. The collaboration is seamless, preventing the need to send links.

Teams can do a SMED project once for every routine they develop and regularly review it. Scrum teams, for example, do retrospective meetings to review their team processes so it's a good time to keep in mind the SMED principles and identify new external tasks. The beauty of a SMED meeting is that you are not only cutting unproductive meeting time, but you are also better ready to solve issues during the day, reduce redundancy, and engage people to communicate more effectively.

Daily Huddles

The Institute for Healthcare Improvement in Boston, Massachusetts[187] has implemented the daily huddles. A huddle is a short, stand-up meeting — 10 minutes or less — that is typically used once at the start of each workday in a clinical setting. In inpatient units, the huddle takes place at the start of each major shift.

In ambulatory surgery centers, huddles occur once per day in each unit (for example, with the operating room staff). In primary care, staff can huddle in the morning to discuss scheduled patients as a team.

The daily huddle gives teams a way to actively manage quality and safety, including a review of important standard work such as checklists. Often, standard work will be the output of previous quality improvement projects, and huddles provide a venue to ensure process improvements stick. Huddles enable teams to review performance and flag concerns proactively.

They also use the situation-background-assessment-recommendation (SBAR)[188] technique, which provides a framework for communication between members of the healthcare team regarding a patient's condition.

- S = Situation (a concise statement of the problem)
- B = Background (pertinent and brief information related to the situation)
- A = Assessment (analysis and considerations of options — what you found/think)
- R = Recommendation (action requested/recommended — what you want)

SBAR is an easy-to-remember, concrete mechanism useful for framing a conversation, allowing expectations to be set for what will be communicated and how.

Agile methods use similar routines to encourage self-organization among team members.

The Scrum Method

The Scrum method is an agile method that includes predefined routines for the team members, which are called Scrum events. As per the Scrum guide developed by Ken Schwaber and Jeff Sutherland, "Prescribed events are used in Scrum to create regularity and to minimize the need for meetings not defined in Scrum. All events are time-boxed events, such that every event has a maximum duration. Failure to include any of these events results in reduced transparency and is a lost opportunity to inspect and adapt."

The events are: sprint, sprint planning, daily Scrum, sprint review, and retrospective.

Sprint is a time-box of one month, a week, or less, a limited time defined by the team in advance to do the work. Projects are divided into sprints.

A sprint meeting is a meeting to define the work to be performed in the sprint.

A daily Scrum is a 15-minute time-boxed event for the development team. The daily Scrum is held every day of the sprint. At it, the development team plans work for the next 24 hours.

A sprint review is a meeting to review what was done during the sprint and what is pending for the next one.

A sprint retrospective is an opportunity for the Scrum team to inspect how the last sprint went with regard to people, relationships, processes, and tools and identify how the team can collaborate better.

Responsibilities are very clear. As mentioned in Skill #10, clear roles are essential to build autonomy. A product owner makes a list of tasks to be accomplished and prioritizes what needs to be done and what can wait. Then the team members choose which tasks to accomplish based on their knowledge, experience, and availability.

All these events are time-box, meaning that the event has a maximum duration to avoid wasting time. To meet the time slot, the tasks during the event should be clear. For example, for the daily

Scrum to be only 15 minutes, the team members who attend need to know in advance what it is going to be discussed at the meeting. For example, the team already defined three questions that every member needs to answer: What did I do yesterday? What will I do today? Do I need any help or have any impediments? With this in mind, everyone is prepared, and meetings flow more quickly.

Methods like Scrum help you stick to a routine of presenting every day what has been done, what needs to be done, what are the problems. Even if you don't do Scrum at work, you can implement some of the principles:

- Define the time to achieve the goal (sprint duration).

- Set a goal to create a potentially shippable product increment every sprint (sprint goal).

- Define actions to accomplish the goal (sprint backlog).

- Define a way to submit requests or help if you can't have daily stand-up meetings.

- Have retrospective meetings to analyze what can be improved in team performance.

Visual Management

When the routines are not clear or there are mistakes, it is usually due to poor team communication. Communication is not always easy; more than 60% of communication takes place nonverbally in a chat with someone, for example. That's why adding a visual part to the verbal part helps to clarify (for example, a picture, a drawing, or a specific color).

Use visual and audible aids to maintain standards and help people comply. Make it easier for everyone around you to detect errors and follow the procedures by clearly exposing what needs to be done with visual cues. It is especially useful when different people have access to the workplace, such as contractors, suppliers, or customers, who are probably unaware of what needs to be done. Help them "see" what they need to do. Even in virtual environments, labels, folder names, and links should help people know what to do first at a glance. If a child can intuitively know what to do, then your process is foolproof.

Key Insight

Use visual and audible aids to maintain standards and help people comply.

The use of boards, physical or online, to help people communicate is widespread. Some companies like paper-based boards: anyone can walk into a team space and see the status of the work. Other companies prefer online boards so everyone around the world can see the status too. Applied mainly in the IT industry, this is a best practice that could be implemented in any industry.

A uniform is a type of visual communication to describe people's roles, not only to other employees but also to the end consumers. When you go shopping and visit a store, uniforms help you identify the manager or salesperson.

You can also apply the 5S methodology (described in Chapter 12) to make the workspace talk in every step and have on hand only what is needed.

Everything can be easily identified by everyone who wants to know what is going on:

- Level of responsibility: Who is in charge of this process? Signs on desks, organizational charts, uniform tags.

- Efficiency: Are we efficient, or do we need to improve? Red, yellow, and green lights on processes or dashboards.

- Storage location: Is this the right place to store this product? Labels and color-coding on drawers and shelves.

- Dangerous places: Can we go in, or should we take precautions? Red-coded doors or floors.

- Status of the process: What is not done yet? Who needs help? "To be done" bins or online folders.

- Runners: Where can I walk? Color-coded floors in green to walk, and red to avoid.

- Emergency exits: Where should I exit? Door signs.

- Scrum boards: What are the tasks that I need to do this week? Three-column boards.

- Kanban cards: What is this product made of? Checklists.

- Satisfaction level: Are our customers happy? Buttons with happy or unhappy faces for customers to push after receiving a service or using a product.

- OK, not OK levels: What is OK, and what is not? Show a picture of how the standard looks for everyone to verify if they are going by the standard. It could be used for products, to detect errors, or even for room organization to ensure cleanliness after use.

These images, colors, and codes are very useful in common areas, such as meeting rooms or hallways. The goal is to be striking, showing users how they should behave. A picture is worth a thousand words.

Routines in Hybrid Teams

How can you visualize team routines when some or all team members work from home or in different locations?

Many companies invest considerably in their offices to create a welcoming place where employees are excited to go and can work effectively. Studies show that well-designed office spaces improve employee well-being and productivity.

This should be no different for home offices. You want your employees to have a place and equipment where they can switch into a work mindset and maximize their productivity. You don't want them in a dark basement, for example.

Give your employees tips on creating a good home office environment that helps them stick to good routines. Encourage them to remove distractions and be organized, which will improve their ability to deliver projects on time.

Even if they don't work from home, but they work in different locations, keeping routines is not easy. Team members may be in different time zones, speak different languages, or celebrate different holiday calendars. Recommend tools or provide training sessions to help acknowledge these differences and agree on common rules and routines to help bridge the culture gap. Use online tools to visualize the routines, from online boards and brainstorming tools, to printouts with reminders and apps that help you drive work. Even the 5S method can be applied in hybrid teams.

Hands-on 11.1

Team Exercise

Think about team routines that are not working well because are causing delaying or mistakes. Discuss with your teammates what is going wrong, and how you can improve it to be more efficient.

_____

_____

_____

References

181. Erik Nereng, "Sunsama's Brilliant Welcome Email Series," *Nereng*, 2020, https://nereng.net/sunsamas-brilliant-welcome-email-series/

182. Michael E. Porter and Nitin Nohria, "How CEOs Manage Time," *Harvard Business Review*, 2018, https://hbr.org/2018/07/how-ceos-manage-time.

183. Ryan Fuller, "A Primer on Measuring Employee Engagement," *Harvard Business Review*, 2014, https://hbr.org/2014/11/a-primer-on-measuring-employee-engagement

184. Ed Catmull with Amy Wallace, *Creativity Inc.* (New York: Random House, 2014).

185. Jared Spataro, "The Future of Work—The Good, the Challenging & the Unknown," *Microsoft*, 2020, https://www.microsoft.com/en-us/microsoft-365/blog/2020/07/08/future-work-good-challenging-unknown/.

186. Ryan Fuller, "Quantify How Much Time Your Company Wastes," *Harvard Business Review*, 2014, https://hbr.org/2014/05/quantify-how-much-time-your-company-wastes.

187. Institute for Healthcare Improvement, "Huddles," http://www.ihi.org/resources/Pages/Tools/Huddles.aspx.

188. Institute for Healthcare Improvement, "SBAR Tool: Situation-Background-Assessment-Recommendation," http://www.ihi.org/resources/Pages/Tools/SBARToolkit.aspx.

12

Skill #12: Build Self-Organizing Habits

I had to free myself from previous ways of working when I was trained to manage and control. I had to let go of that there. The big difference is that, really, I am not responsible, the responsibility lies with the teams.

–An employee at Buurtzorg[189]

F or employees to own and take advantage of the power you want to give them, you have to help them build self-organizing skills.

Self-organizing skills are not new, but they are not widespread either. Usually leaders learn to dictate procedures while employees learn to follow them without asking questions.

Self-organization, instead, is a process in which behaviors emerge when team members interact. Just like in biological systems such as a large school of fish,[190] the organization of the team is not achieved through a leader. Instead, "each fish gathers information about its nearest neighbor and responds accordingly." Have you ever seen how a school of fish swim in a super coordinated flow at high speed, without even touching each other? Have you seen birds and termites doing the same or fireflies turning on their "lights" all at the same time? That is the power of self-organization. It is also called spontaneous order: individuals create order.

If animals and insects can do it, can human teams do it too?

We Culture: Setting the Stage for an Innovative Company

The ability to be part of a We Culture where self-organizing teams prosper will be a key skill to survive and thrive in the future of work.

The idea of self-organizing teams has been promoted by the Deming methods since the 1950s, applied by Japanese companies like Toyota, and has recently been popularized by the Agile Manifesto developed in 2000 applied to build agile teams in the IT sector. One of the 12 principles of the manifesto is that "the best architectures, requirements, and designs emerge from self-organizing teams."

For the school of fish to work, they need basically three things to happen:

1. Define, respect, and continuously repeat very simple rules that apply to all individuals (for instance, maintain a certain safety distance while swimming next to the other fish).

2. Constantly observe and sense the environment and team members to detect any changes, threats, or opportunities.

3. React quickly to adapt to change and improve.

These three components make it easier for organizations to adapt to change, normalize it, and embrace innovation as part of their culture. Self-organizers get things done, improve, and innovate even when nobody is watching. A study conducted by McKinsey[191] found "Companies that ranked higher on managing the impact of the COVID-19 pandemic crisis were also those with agile practices more deeply embedded in their enterprise operating models." Agile teams had the characteristic of being self-organized, which allowed them to continue their work almost seamlessly, while non-agile teams struggled and took longer to be productive.

Individuals were effective working from home because they were used to getting things done with minimal direct supervision.

Key Insight
Self-organizers get things done, improve, and innovate even when nobody is watching.

Front-runner companies are looking forward to becoming agile to stay ahead of the competition and become world-class organizations.

Agile Companies

Companies nowadays are no longer looking forward to just selling; they also need to become the customer favorite tomorrow. When Steve Jobs launched the first iPhone, nobody had ever asked for it or imagined anything similar. With its electric and self-driving cars, Elon Musk took Tesla stock to unimaginable values no other automotive company could ever reach. Mark Zuckerberg, in November 2021, announced the transformation of the social media network into a metaverse space where holograms and teletransportation are the norms.

Changes are so fast that process standardization is not enough—constant change and innovation is part of the process too. The key here is who can adapt faster.

It is not the strongest of the species that survives,
it is the one that is the most adaptable to change,
that lives within the means available and works
co-operatively against common threats.

—Charles Darwin

Lean and Six Sigma process improvement approaches started in the automobile industry and manufacturing in the 1950s, and were successfully applied to other industries. Nevertheless, the software industry that grew significantly during the 1990s realized that these quality practices were not agile enough. They needed to iterate faster to innovate faster. Being the customer favorite became a survival need. They needed to be agile to adapt to changes in the context, to adapt to the new technologies and the new generations of millennials in their workplace. That's how a group of IT experts developed agile thinking in 2001. They applied lean principles and tools to reduce waste and combined them with other idea-generation tools to enable the company to more easily bring innovation to the workplace.

Agile methods have revolutionized the IT industry and enabled it to grow faster than ever. Over the past 20 years, they have greatly increased success rates in software development, improved quality, and speed to market, and have boosted the motivation and productivity of IT teams.

Apple was one of the first companies to achieve innovation through agile iterative customer involvement. Working with manufacturers and being attuned to the customer, Apple integrated customer experience into product design and development.

> *Processes make you more efficient, but innovation comes from people, calling each other at 10:30 at night with new ideas.*
>
> –Steve Jobs

Lately, and especially after the pandemic, agile thinking has been moving from the software industry to other types of organizations.

Some companies are born agile, like Spotify and Apple, with agile rules that enable them to bring innovation in any sector at any time. Others are made agile, like Zappos, Amazon, Airbnb, GE, Saab, P&G, and even banks like Barclays and Citibank are applying agile methods. Some more conservative and traditional industries like oil and gas are starting in the IT sector. As Stephen Denning says in *The Era of Agile*,[192] "Even big, old firms can undertake an agile transformation if they set their minds to it, and stick with it."

Bottom-line Benefits

Everybody resists change at first. Cultural change is even more difficult, as ingrained habits take time to remove. So, why make the leap?

Because agile methods offer a number of major benefits:

- Reduced costs and higher profitability because processes are less bureaucratic, more flexible, and more customer-oriented
- More frequent innovations starting from these self-organizing teams that are closer to the customer

- Higher employee engagement because team members appreciate the autonomy and expertise they develop

- Increased risks awareness, as all the team members are "sensors" of problems and opportunities

Characteristics of Self-Organizing Teams

All the skills and behaviors that I have described throughout this book help build self-organizing teams. When true self-organization happens, people understand exactly what is expected of them, and they have the ability to do what they think is best to get that done. And they do it, even if nobody is checking on them.

In the connection dimension, they learn to communicate better; in the attention dimension, they learn to analyze and understand better what needs to be done; and in the respect dimension (Part 3), they learn to value their strengths and differences, and in this last dimension of empowerment, they learn to get things done on their own, but together.

Some characteristics of self-organizing teams are:

- **Small teams (connection):** Self-organizing teams are usually small, less than 12 people, including a leader. The magic number is usually six. Jeff Bezos uses the "two-pizza team" philosophy, meaning the team should be small enough to be adequately fed by two pizzas, that is six to ten. Google recommends three to six people. Small teams are more productive, innovative, and easier to empower. The number of communication channels required to get work done increases exponentially as the team grows. By keeping the team small you have fewer communication channels. And the best thing is that people are focused 100%, one project at a time.

- **True to company culture (connection):** The team members share a company culture or common behaviors and habits that they value and respect, which helps them make decisions and achieve results aligned with the company's purpose and the We Culture. Zappos recommends, "Hire slowly for culture and fire quickly for culture."

- **Transparent (attention):** Communication is fast and mostly face-to-face or online to ensure no delays or miscommunication. Everything is online if possible. Everyone can get the same type of training. Much more information is available than in traditional companies. Sharing financial information and customer information promotes transparency. For example, some companies share profits with their employees.

- **Quick decision-makers (attention):** A self-organizing team limits how much work they are expected to complete within a certain time frame. This means the team sets the expectations about what they have to complete.

- **Cross-functional or multidisciplinary (respect):** They consist of a mix of people who should have all the skills and knowledge needed to complete valuable work. Members have different skills to be autonomous within the organization, not requiring help or control from other departments.

- **Teams may have their own budget, and define and execute their own hiring, firing, and learning opportunities (respect–employee journey):** Some companies like Buurtzoorg, a nursing organization with 7,000 employees in the Netherlands, have no HR departments.[193]

- **Respectful of others (respect):** Achieving a diverse workforce and obtaining the potential benefits of diversity is not just about recruiting and hiring a diverse team. To fully experience the benefits of diversity, leaders need to create a workplace wherein members of that diverse team feel appreciated and encouraged to share their perspectives.

- **Autonomous (empowerment):** People are encouraged to choose how they will work. They are encouraged to experiment and create new ways of adding value to customers. The fact is that the people actually doing knowledge work usually have the best idea of how to do that knowledge work, given they have the proper experience. So, why not let them do it their way? This way you get faster and more creative decisions that improve customer satisfaction because they are closer to the gemba, where the customer is or where the action take place. There is not much coordination across teams; each team can work in parallel without other team's help.

- **Disciplined (empowerment):** Even though they make lots of decisions about how they work, they follow through with the agreements that were made at the beginning of the team formation. That's why it is important to set clear goals, values, and rules, and follow-up or assist closely when someone is not able to meet them. Almost everyone can work in a self-organized environment, but people are not usually used to working that way, so it may be hard at first. Some employees may feel uncomfortable, so it is OK if they need to leave.

The Role of Agile Leaders In Self-Organizing Teams

So what do self-organized teams look like? Do they have managers?

Self-organized teams have peer commitments more than hierarchical relationships, where the pyramidal structure is no longer needed. But that doesn't mean there is no structure, only that structure varies. There is no top or bottom. Every team may look like a diamond, but the organization is a structure that accommodates those diamonds as needed. Frederic Laloux's[194] research accounts for three structures:

- In parallel: Every team is autonomous and decides the employee experience for their team members with a minimum supporting structure such as HR or finance (examples, consulting, IT, or manufacturing).

- Webs: Teams are autonomous but are connected to each other as steps in a manufacturing process (examples, complex manufacturing companies, refineries, aerospace or automotive).

- Nested: Similar to parallel but with a support structure that specializes in the particular needs of the teams, such as legal issues, HR, or research and development (example, retail store teams of a big chain can be self-organizing while they are assisted by central purchasing and marketing teams).

These teams need to meet regularly to define roles, needs, goals, and purpose so their specific purpose and commitments are aligned and contribute to the organizational goals. These types of organizations need fewer layers between employees and customers.

No team is more important than the other team. A team can have a leader, a representative, or a coordinator, but this person works with the team and has operational tasks:; he or she is more of a doer than a decision-maker.

It doesn't necessarily mean that leaders are no longer needed, but they can become facilitators. By building an environment where team members can interact better, they can get the information they require from their team and act much faster instead of the leader.

Nobody has more power than others. The main difference with traditional organizations is that there is no hierarchy of power.

The leader actually is more agile; it may be just another team member who has the role to facilitate or connect the resources. Zappos, for example, called them *lead links.* But one of the main changes is that leaders won't be the ones making decisions for the team or holding the critical information.

The role of the leader, if there is any, will depend on the needs and maturity of the team. Leaders may have to set the long-term vision, promote interaction across the organization, define priorities, and help unleash team members' ideas to develop new leaders within the team, or maybe they will only have to do the reporting.

Me Culture: Barriers to Self-Organization

Self-organization is like a muscle—it needs to be trained. It requires that every individual perform their job without much supervision.

Usually, the reason why employees don't perform as expected is that they lack the self-discipline to maintain a working schedule or routine.

Discipline is basically doing things naturally, routinely, like a daily habit. Most of our bad habits are due to laziness or lack of willpower. Self-discipline doesn't have to do with talent or intelligence. It actually has a bigger effect on academic performance than intellectual talent.

Employees can set big plans for themselves, but they usually find barriers in their way to accomplishing them. Sometimes there are external issues that employees cannot manage that set them back (customer's reactions, economic turnarounds, traffic, family issues) or sometimes it is just that they are not strong enough to commit to following something no matter what. Again, their willpower muscles have not been trained enough. It is the same reason why they don't exercise or quit smoking. People procrastinate doing new things and are afraid or unsure about the outcome. Fear of failure shows up or questions like "Why should I do this?" And then people just stop there.

Strengthening self-discipline is not totally up to the employees. Companies can do something about it as well. They can build routines that help turn self-discipline into an organizational habit. As Charles Duhigg mentions in his book *The Power of Habit,* "What employees need are clear instructions about how to deal with inflection points." Those inflection points are pains, temptations, or whatever problems they encounter when they are trying to meet the routine. Companies need to work on eliminating those barriers. In their book *Analyzing Performance Problems,* Robert Mager and Peter Pipe say, "Keep in mind that if they (employees) can do it but aren't doing it, there is a reason; only seldom is the reason either a lack of interest or a lack of desire. Most people want to do a good job. When they don't, it is often because of an obstacle in the world around them."

There are many barriers people use to justify why they can't be self-organized. Work on some of them to reduce the likelihood your team will use them.

Barrier 1: "It is not my job." Solution: Build ownership.

The first and most common answer to why a task has not been done is usually "It is not my job." So the first step to empowering people to do the work is splitting the work into small, clear tasks so each team member knows what he or she has to do and is able to do it. They become owners of the tasks, which gives them the power to make decisions. If the owners are unclear, they may give in to the inner temptation of telling themselves that it is someone else's job, so they only focus on what is clearly theirs. Or sometimes good employees get overwhelmed with their tasks and the inaction of other people that they have to cover up. None of these situations are good or fair. Giving employees more autonomy in the way they work improves self-discipline.

Barrier 2: "I did it as usual" or "I didn't know how to do it." Solution: Provide clear instructions.

Create clear instructions for whatever needs to be done. I have seen "clear instructions" that simply say: mix well, keep clean, or maintain. What is that supposed to mean? How do we know it's well mixed? How do we know it's clean? The answer can be different from employee to employee, so if there is room for a personal perspective or common sense, you may get different outcomes from the same instruction.

The task may have been performed for years by the same person, and he or she may have changed the steps over time without even realizing it. Or the task may have been transferred to different employees through time, with no written instruction, only observations, bringing changes and shortcuts with every change in control.

If an instruction is not being followed correctly, it is probably not written correctly. Use checklists, pictures with standards, labels, color coding, or any other visual management techniques to help the employees perform as expected.

Barrier 3: "It was just a human error." Solution: Understand why.

Every time there is an error, perform a root cause analysis, don't just assume it was a human error. Many investigations are closed just saying it was a human error, such as procedures were not followed, or there was a lack of attention or a human mistake. "Human errors cannot be eliminated by simply telling operators to be more careful. Human error is more a symptom than a cause. Do not use human error as a root cause. Always ask why the human made a mistake."[195] It may be again that the instructions were not clear, but it also could be that the computer was not working properly so the employees looked for a shortcut. Maybe instructions were so long that employees got lost in the explanation. Perhaps the workplace was too noisy or uncomfortable in a way that didn't allow the employees to meet the requirements. Redesign the workplace to make it "error free" by using more visual cues.

Barrier 4: "I thought it was not as important."
Solution: Schedule periodic reviews.

When employees feel that something is not being controlled or supervised, they think it is not important, so they may stop doing it. Reviews and audits are a great way to show that following procedures is important. Analyze together if things are being done the right way, how often, and how effectively. The most important part of an audit is sharing the results so employees can celebrate and get rewarded or discuss what still needs to be changed. Be candid, that is, be clear and specific about what is wrong and, if possible, how it can be fixed or what is expected the next time.

Barrier 5: "I am already doing it."
Solution: Verify with a Socratic walk.

Remember that everything is relative. Sometimes employees say they are doing something, but they are not actually doing it, or they are doing it partially (and this not intentional). The best option is to see it yourself and ask questions. Go to the place where things are happening (the gemba). Go and see what your employees are doing and how, and ask questions: Where are the results of the last audit? Which instruction are you following? Why are you doing this? Could you do it differently? What are the risks and what are the improvements in progress? Both you and the employee can learn a lot from this. The power of the Socratic walk is that leaders are not there to correct the employees or tell them how to do things. They are just asking questions that may help them understand the employees' problems, or they may help the employees see their own work differently. They may even realize something is wrong, but they will feel like they found it by themselves. This way, we nurture the continuous improvement culture habit without hurting any feelings. Plan for Socratic walks and include them in your weekly routine.

Barrier 6: "It was not my fault." Solution: Look for inflection points and be prepared to prevent them.

You cannot control everything that may happen to your employees, but for sure inflection points will be the moment when the employees will have the temptation to forget or twist the rules, or just say "it was not my fault."

These inflections could happen when a customer gets frustrated, when your team encounters a high-risk situation, or when you are in a discussion with incomplete data. Inflection points vary based on the job. You need to collect all the behaviors that your team believes are usually incorrectly handled and provide guidance on "the right way" to deal with them. You can even make employees think of how to handle them in the future. These desired behaviors can be practiced doing role-playing so new behaviors easily become part of their daily routines. If they repeat and repeat, when they actually find themselves in the situation, they will be better prepared to handle it. Once they implement it, the rewards of getting things done the right way will help them keep on the right track.

> *As willpower muscles strengthened, good habits seemed to spill into other parts of their lives.*
>
> –Charles Duhigg

All these barriers are commonly seen in all types of industries, but there is a way out. Self-discipline can be learned and practiced. You need to build an environment in which self-discipline is not as painful. The great thing is that good habits then can be applied not only at work but also in other parts of our lives.

Barrier 7: "It was their fault." Solution: Coach the bad apples.

There is always a bad apple, someone who repeatedly mimics Me Culture habits and corrodes the team's efforts. There is no bad apple that can last forever. Identify them, empathize with them, and coach them to do better. A great partner or leader can always help a bad apple to do better too. They must have a reason for how they behave: a previous experience, sense of unfairness/onlyness, or feeling left out. Understand the reasoning and help drive the energy to a fruitful team purpose.

Once the bad apple changes behaviors, the team will flow much faster in the right direction. What's more, bad apples are so influential that they get copied by everyone no matter if they do good or bad. Once you convert their *me* habits into *we* habits, they will turn into "we ambassadors." Don't miss the opportunity to get them on board!

Can My Team Build Self-Organizing Skills?

You probably have a lot of questions regarding self-organizing teams, such as: How do you build them? Is any type of organization better at creating them? How long does it take to turn the whole organization into an agile self-organizing setting?

The cultural transformation can start in a pilot group or with all teams at once. It is usually recommended to start with a pilot "we team." Once the first team has started and there are some learnings on what worked and what did not, then you can continue with the next teams in waves.

To choose where to start, leaders must consider multiple criteria, including strategic importance, budget limitations, availability of people, return on investment, cost of delays, risk levels, and interdependencies among teams. Some companies choose based on pain points felt by customers, and others start with the more mature teams to make sure the pilot is a success.

It is important to discuss some considerations before deciding where to start:

- Any organization can become self-organized, but some teams will have it easier than others, depending on the types of tasks they perform, the prevailing culture, and the leaders who guide the effort.

- Not every function needs to be organized as a self-organizing team, though a "we culture approach" company-wide is desired to ease the work across groups. That means even though the structure of some of the functions may not change, every employee should start getting the "we culture" habits in terms of systems and teamwork, caring for internal and external clients, embracing empowerment, and practicing self-discipline.

- Routine operations such as purchasing and accounting are less fertile ground for self-organizing teams, as they are used to working in silos. But they still can be achieved and provide lots of benefits to increase flow, efficiency, and employee engagement. Zappos has applied the system company-wide. Other growing organizations like Netflix, Spotify, Ideo, and Apple have also applied self-organizing teams successfully. On my side, I have applied this culture to companies in industrial settings, manufacturing, retail, and even in non-profits. Hospitals and schools also benefit from them. Even families can start self-organization at home.

- Agile and self-organizing teams support the ones that are not self-organized. So once you begin launching these teams, you can't just leave the other parts of the business alone. Some changes may be needed, or again, people should be trained on the new habits to be able to understand and serve the other groups.

- Leaders need to be aligned and in agreement with the new setting, as they are the example that everybody will follow. I usually find that the main constraints are middle leaders. Middle leaders are pivotal in this change, so they need special attention. Usually, they are evaluated on how well they control their teams, but in a self-organizing team, command and control are to be avoided. For them, at the beginning is a big mindset change that needs to be acknowledged and addressed individually and as a team.

- When teams need to contact other departments such as budgeting, hiring, and procurement to get approvals or make decisions, they drive delays and lack flexibility. Offer to have teams do these tasks themselves instead of outsourcing them to other departments. For example, self-organizing teams can choose to manage their budgets or hire their personnel.

We Culture Tool for Agility and Self-Organization: The 5S Methodology

I found the 5S methodology to be very helpful in driving self-organization and developing team routines.

The 5S methodology is part of the lean toolkit; 5S is a set of habits that drive individual productivity and team self-organization through a lean workplace. Employees learn to become more engaged, autonomous, and innovative in every endeavor, even working from home.

Through five steps—sort, store, shine, standardize, and self-organize—team members learn what to do, how to do it, and when to do it without being told, just by observing and sensing the environment. If everyone in the organization is trained in 5S, it is much easier to highlight and eliminate waste, find tools, and visualize what needs to be done.

More than a method, 5S implies a change in the way you work, a cultural change. It is one of the first tools recommended for a company that is trying to reduce unneeded costs, increase employee engagement, and improve quality, productivity, and safety at the same time.

The 5S method has its origins in Japan. It was first popularized by Taiichi Ohno, who designed the Toyota Production System, and Shigeo Shingo, who also put forward the concept of poka-yoke, a technique to prevent errors.

The five steps are based on five Japanese terms, all starting with "S" (the English version also starts with "S"), which represent steps toward the final aim, which is team self-organization.

1. **Sort:** Separate unneeded items to increase focus on what is needed.

2. **Store:** Organize all the needed items in a specific "home" or pre-identified and labeled location to have them ready to use and available to anyone who might need them.

3. **Shine:** Set a new level of cleanliness to maintain the needed items in good condition and be able to spot potential problems quickly.

4. **Standardize:** Find a simple system to sort, store, and shine while you work

5. **Self-organize:** Repeat the 5S steps every day until they become a habit, observe any changes in the environment, and improve as needed.

By practicing the 5S every day, individuals start new habits for themselves and build an environment that helps them and their teammates sustain these new habits in the future. By sorting what is not needed, labeling and storing what is needed, and promoting cleanliness, the cues to tell you what to do and when are more evident, keeping you focused on what is essential and urgent. We have many sensors in our bodies: taste, smell, touch, vision, hearing. 5S urges team members to use all the senses to identify clues about what they need to focus on too, from colorful labels, alarms, and music, to touching surfaces and smelling weird odors as part of control checks.

Employees start to understand how the workplace impacts their moods, behaviors, and successes, so they learn to take care of it as an integral part of their day-to-day jobs. Interactions within and across teams are based, just like the fish, on signals and cues derived from the environment, work in progress, and fellow workers. Combined with the 5S steps, these signals are simple enough to allow team members to know what to do and adapt quickly through small changes or micro changes.

In a family, cleaning and keeping the house organized can be a one-person show, or the entire family, including the kids, can be invited to participate. When all the family members understand what it takes to keep the house clean, they all engage actively not only to clean but also to avoid getting things dirty. 5S helps the "family" self-organize and arrange items purposefully so everyone knows what to do, how, and when without being told.

The last step of 5S is to sustain the self-organizing skills and habits over time. Just like the Shu Ha Ri explained in Skill #9, sustaining self-organization is all about defining and repeating the 5S rules, observing the environment and teammates for cues, and proposing improvements when needed (see Figure C12.1).

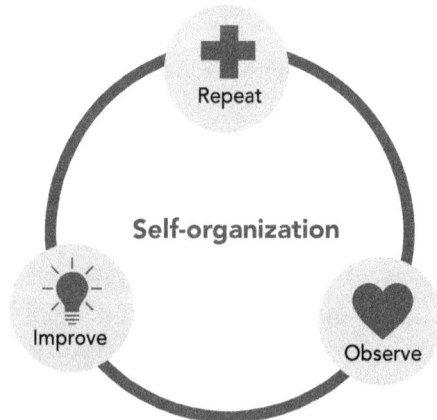

Figure C12.1   Self-organizing culture cycle.

Hands-on 12.1

Check your workplace. How can it help you build self-organizing skills? Fill in the list with improvement opportunities. List the items you don't need that distract you and the items you need but are hard to find, and commit to making at least one micro change to your behavior that sticks for the remainder of the year.

Individual Exercise

Check around your workplace. How can it help you build self-organizing skills? Fill in the list with at least three improvement opportunities.

Items you don't need.	Items you need but don't find easily.	Commit to one micro change this week.

Summary

Achieve agility by sharing the power to sense change and react quickly.

Recap 1
Promote autonomy.

Recap 2
Encourage team routines.

Recap 3
Build self-organizing habits.

Reflection Time

Take five minutes to think about three highlights from the dimension of EMPOWERMENT. Write them on your note pad or the action plan available on the We Culture app.

Three Highlights

_____

_____

_____

References

189. Frederic Laloux, *Reinventing Organizations* (Brussels, Belgium: Nelson Parker, 2014).

190. Tony Hsieh, *Delivering Happiness* (New York: Grand Central Publishing, 2013), 197.

191. Scott Camazine, Jean-Louis Deneubourg, Nigel R. Franks, James Sneyd, Guy Theraulaz, and Eric Bonabeau, "Self-Organization in Biological systems" (Princeton, NJ: Princeton University Press, 2001).

192. https://www.mckinsey.com/business-functions/people-and-organizational-performance/our-insights/an-operating-model-for-the-next-normal-lessons-from-agile-organizations-in-the-crisis

193. Stephen Denning, *The Era of Agile* (New York: Amacom, 2018).

194. Frederic Laloux, *Reinventing Organizations* (Brussels, Belgium: Nelson Parker, 2014).

195. *QP*, Sept 2018, https://asq.org/quality-progress/articles/pitfalls-and-prat falls?id=e6158f3aa9974387bcdfaa58f45571a6.

Conclusion:
NEXT STEPS

W hile anyone in the organization can practice these 12 CARE behaviors regardless of their role, hierarchy, or experience, and propose changes (bottom-up approach), if you are looking for an organization-wide cultural change, the effort also requires top-down involvement. Remember that as the "we culture thinker" you have now become, there should be no top or bottom, but change needs to be organized intentionally to succeed.

Change is not easy. As we reviewed in Chapter 3, change management programs have a high failure rate of about 70% because management and employees are not aligned.[196]

The change management process should embed the 7Rs (see Chapter 1) and the 12 CARE supporting behaviors into the entire employee experience (Skill #9) to reinforce the desired culture.

By aligning people with the desired culture, everyone understands what needs to change, has incentives to do so, and follows models.

The next question is probably: Where do we start? The cultural change model M3 will guide you through the main steps to start your journey to develop an intentional culture based on your company's values, goals, and purpose.

During the mapping stage, you will define your north star if you haven't done it yet; that is, where you want to be, what is your gap, and what is the action plan to implement to manage the change stage.

What to prioritize will come from your culture survey and gap analysis, but it also has to come from your inner desire to be more intentional on a specific dimension of the CARE spectrum.

Every person, as well as every organization, is born to shine on one of the four CARE dimensions. The dimensions represent the different parts of the body of the organization (or the individual): the connection is the soul and intuition (the spiritual side), the accountability is the brain (the rational side), the respect is the heart (the emotional side), and the empowerment is the arms and the legs (the physical side).

While organizations need to focus on all four dimensions to be balanced and successful in the long term, they will indefectibly (and fortunately) be better at one of them.

That is why even looking forward to becoming we cultures, there will be different subtypes of cultures, based on the main goals and strengths of the organization:

- Focus on connection will primarily drive innovation

- Focus on attention will primarily drive quality

- Focus on respect will primarily drive employee engagement

- Focus on empowerment will primarily drive agility

Discover what your focus is as an organization and as a person by analyzing which skills you are best at, and continue developing them. Perform a personality test for yourself and your teammates to find out your CARE profile. Again, you will also need to develop the rest of the skills, but you may need some cooperation from other organizations or individuals who excel at them. We cultures will always look to be better at what they do best and complement their skills with others to find a delicate balance that is healthier and more productive for everyone. Collaboration will be the core to achieving the greater good.

. . . .

Find out your CARE self-assessment at the
We Culture App. Download it from your
app store or at www.theweculture.com.

. . . .

For example, if your company is in the technology industry and develops apps, you are probably focusing on innovation. You need to develop the skills of fostering change, networking, and being driven by purpose first. For innovation to be part of everybody's job, you need to build a context that makes team members and leaders feel comfortable coming up with new ideas, encourage employees to feel free to share them, and find a team to implement them. You will need to retrain your team's brains to learn new behaviors. But in order to get the best out of these skills, you need to work on employee engagement. While innovation may come easy for you initially (it is your strength after all), you also need engaged employees and teamwork, so you will probably require some research and coaching on how to develop this area.

Likewise, within the company and across teams, different skill sets, personalities, and preferences will need to be combined in diverse and multidisciplinary teams to help unleash the best-balanced and synergic teams.

The emphasis on each dimension can also vary based on which stage your organization is going through (see Figure C13.1).

Figure C13.1   CARE stages.

At the formation of the organization, the connection skills may be more important to develop new skills, design new products, or find a team. For mature companies that are trying to maintain their performance, their focus on attention and respect may be crucial. To do a productivity leap might be a concern for star companies when focus on empowerment can drive the best results utilizing fewer resources. But these stages are nothing but dynamic, so the organization may be focusing on respect while finding that it needs to redesign its values or reimagine its purpose. Results can be frustrating, highlighting a crisis and the importance of redesigning processes and team routines continuously.

Likewise, each individual will also be moving from connection to attention, and going through different stages and preferences in their own lives. When the stage matches your skills, you are at your best. No journey is the same, and there is no single way to go around it. But turning everyone into sensors of change helps to move faster and react better.

For instance, you are always moving forward full of energy (your inner focus is empowerment), and you happen to be at this stage of your life where you are moving forward, going a direction you enjoy, and achieving your desired goals. But you can also be moving to the attention stage, where you are looking at your results, and they are not going as expected. You are clearly not comfortable here; you don't know what to do. You keep moving, as it is in your instincts but it doesn't seem like you're going anywhere. It is a moment of crisis where you will have to work hard to get back to your comfort zone. Change is constant, so you will always have to work hard: both when you are in "your zone," because you love it, and also when you are out of it because you want to escape.

Start mapping the change by doing personality tests for the team and a culture survey for the company as a whole, identify the main pain points that impact your ROI, and continue the journey until everybody is empowered to drive a We Culture (see Figure C13.2).

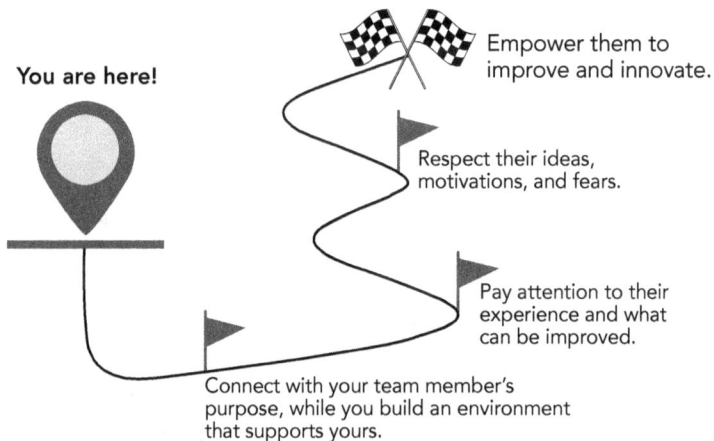

You are here!

Empower them to improve and innovate.

Respect their ideas, motivations, and fears.

Pay attention to their experience and what can be improved.

Connect with your team member's purpose, while you build an environment that supports yours.

Figure C13.2   CARE journey.

Organizational Change or Individual Change?

The hard nut to crack is how to make the change stick to the whole organization. A "me" organization will find it difficult to change at first; the focus on avoiding errors, the need for structure, and the focus of individuals on survival makes it hard to be agile to change.

We cultures instead accept change as part of the journey and enjoy it. They change all the time. Structures are more malleable, as everybody contributes to promoting changes, even clients and suppliers. Everybody is responsible, not guilty. We organizations lead processes through meaningful interactions among their members and turning everyone into organizational sensors. Therefore, change has to be both individual and organizational to be successful.

Key Insight

We organizations lead processes through meaningful interactions among their members and turning everyone into organizational sensors.

Transforming a culture is a matter of identifying the behaviors you would like to see become consistent practices and then instilling the discipline of actually doing them through the models that this book shares.

Shifting to an agile structure where the We Culture thrives takes time and discipline. Some organizations will start slow, while some will make a big leap. Once the first team has started and there are some learnings on what worked and what did not, the new routines will spread like wildfire.

W. Edwards Deming would say that there are usually two main challenges to achieving change: Why change if things are good? And why would I help or listen to you? So, why change to "we" if we are doing "good" this way? Because there are various degrees of good. You could be doing better. All lawyers are supposed to be certified to win cases, but are they all winning cases? In the same way, there are different types of lawyers, different grades of doctors, and different grades of companies. Yours is probably profitable. Could it be even more profitable?

And why help or listen to others while "I know how to do my job"? Because jobs are never totally separate from each other. Everything you do, even if you're a solopreneur, influences some-one: another partner, a client, a supplier, or a teammate. So if you listen to others or ask them how your job impacts their jobs, you will get a different perspective that you cannot get working alone.

Get the help of an external consultant if getting started is too challenging. It's not always possible to see change within your organization; sometimes, you need someone from outside to help you see things differently. But never hesitate, do it. Start your We Culture journey today!

I hope you enjoy it!

· ·

Dream big, never assume there is nothing you can do about a problem. Divide it into small pieces and focus all in one by one. Feel safe that nobody will judge you and everything will be ok. If not, simply learn from it.

–Luciana Paulise

· ·

Summary

CARE Values of the We Culture

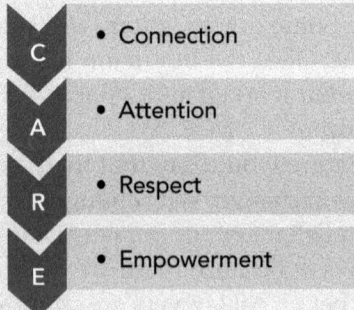

- C • Connection
- A • Attention
- R • Respect
- E • Empowerment

References

196. McKinsey & Company, "Why Do Most Transformations Fail? A Conversation with Harry Robinson," 2019, https://www.mckinsey.com/~/media/McKinsey/Business%20Functions/Transformation/Our%20Insights/Why%20do%20most%20transformations%20fail%20A%20conversation%20with%20Harry%20Robinson/Why-do-most-transformations-fail-a-conversation-with-Harry-Robinson.pdf.

Appendix A
CARE Model Background: Where It All Started

M y interest in a We Culture model of desired behaviors started long ago, when I was finishing my bachelor's degree in business in Argentina at the Morón University and had to work on a thesis on leadership. I compared the main leadership authors I had read at that time and realized how they all had one thing in common: all their frameworks could be analyzed in terms of four dimensions of the human body: the spiritual, the mental, the emotional, and the physical discipline of the leader. See Table A.1.

James Hunter	Warren Bennis	Stephen Covey	Daniel Goleman	Deming	Dimension
Authority	Vision Curiosity	Vision	Vision Strategy	Appreciation for a system	Intuitive CONNECTION
Service	Integrity	Consciousness	Confidence Influence	Knowledge of variation	Rational ATTENTION
Love	Passion	Passion	Empathy	Psychology	Emotional RESPECT
Sacrifice Will	Audacity Trust	Discipline	Achievement Commitment	Theory of knowledge	Physical EMPOWERMENT

Table A.1 Different frameworks that aligned with the CARE model.

The most revolutionary and applicable to my thinking was W. Edwards Deming's approach to management. As a quality engineer, I became interested in learning more about his thoughts. That is how in 2014 I got a scholarship from the Deming Institute to visit the Library of Congress in Washington DC to access material that belonged to him and was not even published. In the book, you may have noticed many notes and paragraphs extracted from the manuscripts I found at the Library of Congress. If you want to take a look at them, go to the app We Culture or visit www.theweculture.com to find pictures of the original manuscripts.

. . . .

Find Deming Library of Congress manuscripts
by clicking on More about Deming at the
We Culture App or visit www.theweculture.com.

. . . .

One of his latest contributions was the system of profound knowledge (SoPK). The SoPK was first introduced in his book, *The New Economics* in 1993.[197] The four areas of the system are:

1. Appreciation for a system

2. Knowledge variation

3. Psychology

4. Theory of knowledge

1. A system view helps create a long-term focus. Rather than seeing incidents as isolated (and often looking for the person to blame for a bad result), a system view allows managers to connect the incidents and the results, so as to find systemic long-term solutions. For Deming, the purpose of an organization was to create a system that provides benefits to all stakeholders. Taking a systems approach resulted in viewing the organization in terms of many internal and external interrelated connections and interactions, as opposed to discrete and independent departments following the chain of command.

   Appreciation for a system is aligned with the connection dimension.

2. Knowledge variation: While Deming is most known for using data to improve the performance of organizations through the PDSA and control charts, he emphasized the need to train workers to understand the data. He was against managing by the numbers.

   Knowledge of variation is aligned with the attention dimension.

3. Psychology: Deming acknowledged, even when most companies were aligned to the Taylorist approach to people, that people were fundamental to improving and making the organization successful. He considered that they didn't simply have to obey, but instead, they had to be set free to offer ideas and foster their pride at work. They should be motivated intrinsically to work better, not just externally through monetary rewards and competition. Actually, he proposed that the job of a leader was not to motivate people, but to remove the barriers to joy in work by creating an environment where people could take pride in what they do.

   Psychology is aligned with the respect dimension.

4. Theory of knowledge: The last step is the theory of knowledge, which is to gather evidence on real-life experiments to support the theories. It is the hands-on stage. The model used within the Deming system for managing to gain evidence was the PDSA, which with a cycle of continuous planning, doing, studying, and acting, employees would implement ideas and adjust, assuming that change was the constant.

   Theory of knowledge is aligned with the empowerment dimension.

References

197. W. Edwards Deming, *The New Economics: For Industry, Government, Education,* second edition (Cambridge, MA: The MIT Press, 1994), 50.

NOTES

NOTES

NOTES

NOTES